Lecture Notes in Mathematics

A collection of informal reports and seminars
Edited by A. Dold, Heidelberg and B. Eckmann, Zürich

Series: Mathematisches Institut der Universität
Erlangen-Nürnberg. Advisers: H. Bauer and K. Jacobs

121

AF060870

Herbert S. Bear

University of Hawaii, Dept. of Mathematics, Honolulu, HI/USA

Lectures on Gleason Parts

Springer-Verlag
Berlin · Heidelberg · New York 1970

This work is subject to copyright. All rights are reserved, whether the whole or part of the material is concerned, specifically those of translation, reprinting, re-use of illustrations, broadcasting, reproduction by photocopying machine or similar means, and storage in data banks.

Under § 54 of the German Copyright Law where copies are made for other than private use, a fee is payable to the publisher, the amount of the fee to be determined by agreement with the publisher.

© by Springer-Verlag Berlin · Heidelberg 1970. Library of Congress Catalog Card Number 74-114555. Printed in Germany.
Title No. 3277.

CONTENTS

Section

0. STANDING ASSUMPTIONS AND NOTATIONS

These lectures are concerned with complex function algebras, and real function spaces. We will give here the basic definitions and notations we will use throughout.

Let X be a compact Hausdorf space. $C(X)$, and $C_R(X)$ will denote the spaces of all continuous complex valued, and real valued, functions on X. We always use the sup-norm in $C(X)$, $C_R(X)$, and subsets of these spaces: $\|f\| = \sup\{|f(x)|: x \in S\}$. A <u>function algebra on</u> X is a closed subalgebra A of $C(X)$, which contains the constant functions and separates the points of X. A <u>function space on</u> X is a linear subspace B of $C_R(X)$, which contains the constant functions and separates the points of X. We will frequently be concerned with the situation where A is an algebra on X and $B = \mathrm{Re}\, A$.

If A is an algebra on X, then there is a closed set Γ with the property that every function in A assumes its maximum modulus on Γ, and Γ is a subset of every closed set with this property. The same statement holds for a function space B. This set is called the Silov boundary of A or B in X. If $B = \mathrm{Re}\, A$, then A and B have the same Silov boundary.

The <u>spectrum</u> S_A of a function algebra A is the set of all $(\neq 0)$ multiplicative linear functionals on A. These homomorphisms are automatically continuous, so $S_A \subset A'$ = the space of all norm continuous linear functionals. We give S_A the w*-topology of A' = the $w(A', A)$ topology = the topology of pointwise convergence on A. Then S_A is compact, and the map $x \to e_x$ (where $e_x(f) = f(x)$ for $f \in A$, $x \in X$) is a homeomorphism of X into S_A. We can extend the functions $f \in A$ to S_A by writing $f(\phi) = \phi(f)$ for $\phi \in S_A$. Then the Silov boundary is the smallest possible compact set on which we can represent A, and S_A is the largest. The Silov boundary of A is the same set, whether A is considered as an algebra on S_A or on some proper closed subset X of S_A.

For a function space B, we consider the positive linear functionals of norm one in B': $T_B = \{F \in B': F(1) = \|F\| = 1\}$. Evaluation maps X homeomorphically into a subset of T_B, with the w*-topology = $w(B', B)$ topology.

The carrier space T_B is the largest space on which the functions of the linear space B can be extended with constants going into constants, and the norms preserved. T_B is the closed convex hull in B' of the image of X under the evaluation embedding. The Silov boundary of B in X or T_B is mapped into the closure of the set of extreme points of T_B. The linear space B, represented on T_B, becomes the set of all restrictions - to - T_B of w*-continuous linear functionals on B'.

If A is a function algebra, and $f \in A$, $\|f\| < 1$, then $(1 - f)^{-1} = \sum_0^\infty f^n$, since A is assumed to be closed. Consequently if ϕ is any fractional linear transformation,

$$\phi(z) = \frac{z - \alpha}{1 - \bar{\alpha}z},$$

with $|\alpha| < 1$, then $\phi \circ f \in A$ for all $f \in A$ with $\|f\| < 1$. It is also true that if ϕ is any function analytic on a neighborhood of $f(S_A)$, then $\phi \circ f \in A$, but we shall not need this result.

1. EXAMPLES OF GLEASON PARTS

One of the persistent goals in the study of function algebras is to find conditions under which there is "analytic structure" in the spectrum. For example, one hopes to find maps ϕ: $D \to S_A$, where $D = \{z: |z| < 1\}$, such that $f \circ \phi$ is analytic on D for all $f \in A$. The consequences of such a map which follow from the Schwarz lemma led to the idea of a Gleason part.

Definition 1-1. Let A be a function algebra on X, and let B be a function space on X. For points $x, y \in X$, let

$$G(x, y) = \sup\{|f(x) - f(y)|: f \in A, \|f\| < 1\}$$

$$\sigma(x, y) = \sup\{|f(y)|: f \in A, \|f\| < 1, f(x) = 0\}$$

$$d(x, y) = \sup\{|\log u(x) - \log u(y)|: u \in B, u > 0\}$$

Clearly $G(x, y) \leq 2$, $\sigma(x, y) \leq 1$, and $d(x, y) \leq \infty$. G and d are obvious metrics on X. Thinking of the points of X as functions respectively on the unit ball of A , or the set of $\log u$ for positive $u \in B$, then G and d describe the (norm) topology of uniform convergence on these respective sets. We will show that σ is also a metric on X , and that if $B = \mathrm{Re}\, A$,

$$(1\text{-}1) \qquad d(x, y) = \log \frac{1 + \sigma(x, y)}{1 - \sigma(x, y)} = 2 \log \frac{2 + G(x, y)}{2 - G(x, y)} .$$

The condition $d(x, y) < \infty$ clearly defines an equivalence relation on X , which we denote $x \sim y$. Since $d(x, y) < \infty$ iff $\sigma(x, y) < 1$ iff $G(x, y) < 2$, (still assuming that $B = \mathrm{Re}\, A$) these latter conditions also define an equivalence relation on X . The equivalence classes of X are called the Gleason parts of X . The equation (1-1) also shows that d, G, and σ define the same topology on each Gleason part of X .

Before we prove (1-1), we will consider several examples, temporarily assuming the identity (1-1). The following lemma allows one to identify some trivial (singleton) parts.

Lemma 1-2. If x is a peak point for an algebra A , then $\{x\}$ is a part.

Proof. A peak point x is a point such that for some $f \in A$, $|f(y)| < |f(x)|$ for all $y \neq x$. We may assume that $f(x) = \|f\| = 1$. For fixed $y \neq x$, $f^n(x) = 1$, and $f^n(y) \to 0$. It follows (consider $f^n - f^n(y)$ divided by its norm) that $\sigma(y, x) = 1$, so that y is not equivalent to x .

Example 1. Let X be the closed unit disc, $X = \bar{D} = \{z: |z| \leq 1\}$, and let A be the algebra of all continuous functions on $\bar{D}$ which are analytic on D . This algebra is called the disc algebra. The spectrum of the disc algebra is $\bar{D}$, and the open disc D is one part. Each point on the circle $\{z: |z| = 1\}$ is a singleton part. The Schwarz lemma says exactly that $\sigma(0, z) = |z|$ for each $z \in D$ ($|z| < 1$), so that D is the part of all points equivalent to 0 . Each point of the circle is clearly a peak point.

Example 2. Consider two tangent discs $X = \{z: |z| \leq 1\} \cup \{z: |z - 2| \leq 1$, and let A be all continuous functions on X which are analytic on the interior. Each open disc is contained in a part by the Schwarz lemma, and each boundary point is a peak point. To see that the open discs are separate parts, let $f(z) = z$ for $|z| < 1$, and $f(z) \equiv 1$ for $|z - 2| \leq 1$. Then $\sigma(0, z) = 1$ for all z with $|z - 2| < 1$.

Example 3. (The cylinder algebra). Let $X = \overline{D} \times [0, 1]$ and let A be all continuous functions $f(z, t)$ on X such that $f(\cdot, t)$ is analytic for $|z| < 1$ for each fixed $t \in [0, 1]$. Each horizontal open disc $D \times \{t_0\}$ is contained in a part. Different discs are distinct parts, as one can see by considering a real continuous function which is constant on each slice, and 0 on one slice and 1 on another. Again boundary points are peak points, and therefore are singleton parts.

Example 4. (Cylinder algebra with centers identified). Let A be the subalgebra of Example 3 consisting of all $f(z, t)$ such that $f(0, t)$ is a constant. This is the same as identifying $\{0\} \times [0, 1]$ as a single point. Then there is only one non-trivial part, consisting of the union of the open discs (with centers identified).

Example 5. (Bicylinder algebra). Let X be $\overline{D} \times \overline{D}$, and let A be all continuous functions $f(z, w)$ which are analytic in z, for $z \in D$, for each $w \in \overline{D}$, and vice versa. Then $D \times D$ is one part, since $(z_1, w_1) \sim (z_1, w_2) \sim (z_2, w_2)$ if $|z_i| < 1$, $|w_i| < 1$. If $|z_0| = 1$, then $\{z_0\} \times D$ is a part, and similarly $D \times \{w_0\}$ is a part if $|w_0| = 1$. Each point $\{(z_0, w_0)\}$, $|z_0| = |w_0| = 1$, is a trivial part.

2. PROOF OF THE d-σ-G IDENTITY

We proceed to prove the identity

$$d(x, y) = \log \frac{1 + \sigma(x, y)}{1 - \sigma(x, y)} = 2 \log \frac{2 + G(x, y)}{2 - G(x, y)} \tag{2-1}$$

We assume that G and σ are defined by some function algebra A on X, and that d is defined by the function space $B = \operatorname{Re} A$.

Lemma 2-1.

(i) If $|\alpha| < 1$, $|\beta| \leq 1$, then $|\alpha|^2 + |\beta|^2 \leq 1 + |\alpha|^2|\beta|^2$.

(ii) If $|\alpha| < 1$, then $\left|\frac{\alpha - \beta}{1 - \bar{\alpha}\beta}\right| \leq \frac{|\alpha| + |\beta|}{1 + |\alpha|\,|\beta|}$.

Proof.

(i) $$\begin{aligned} |\alpha|^2 + |\beta|^2 &= |\alpha|^2 + |\alpha|^2|\beta|^2 + (1 - |\alpha|^2)|\beta|^2 \\ &\leq |\alpha|^2 + |\alpha|^2|\beta|^2 + 1 - |\alpha|^2 \\ &= 1 + |\alpha|^2|\beta|^2 . \end{aligned}$$

(ii) If $B \geq A$, then $(A + t)/(B + t)$ is increasing. With $A = |\alpha|^2 + |\beta|^2$ and $B = 1 + |\alpha|^2|\beta|^2$, we have therefore

$$\begin{aligned} \left|\frac{\alpha - \beta}{1 - \bar{\alpha}\beta}\right|^2 &= \frac{|\alpha|^2 + |\beta|^2 - 2 \operatorname{Re} \alpha\bar{\beta}}{1 + |\alpha|^2|\beta|^2 - 2 \operatorname{Re} \alpha\bar{\beta}} \\ &\leq \frac{|\alpha|^2 + |\beta|^2 + 2|\alpha|\,|\beta|}{1 + |\alpha|^2|\beta|^2 + 2|\alpha|\,|\beta|} \\ &= \frac{(|\alpha| + |\beta|)^2}{(1 + |\alpha|\,|\beta|)^2} \end{aligned}$$

Lemma 2-2. If $f \in A$, $\|f\| < 1$, then

(2-2) $$\frac{|f(x) - f(y)|}{|1 - \bar{f}(x)f(y)|} \leq \sigma(x, y) .$$

Proof. Let $g = (f(x) - f)/(1 - \overline{f(x)}f)$. Then $g \in A$, $\|g\| < 1$, and $g(x) = 0$. Therefore $g(y) \leq \sigma(x, y)$. Since $g(y)$ is the left side of (2-2), we are done.

Lemma 2-3. $\sigma(x, y) = \sigma(y, x)$.

Proof. If $f \in A$, $\|f\| < 1$, $f(y) = 0$, then (2-2) shows that $|f(x)| \leq \sigma(x, y)$. That is $\sigma(y, x) \leq \sigma(x, y)$, and consequently $\sigma(y, x) = \sigma(x, y)$.

Lemma 2-4. If $f \in A$, $\|f\| < 1$, then

(2-3) $$|f(y)| \leq \frac{\sigma(x, y) + |f(x)|}{1 + \sigma(x, y)|f(x)|} .$$

Proof. Let $f(x) = \alpha$, and $\beta = (f(x) - f(y))/(1 - \overline{f(x)}\, f(y))$. Then $|\beta| \leq \sigma(x, y)$ by (2-2), and computation shows that $(\alpha - \beta)/(1 - \bar{\alpha}\beta) = f(y)$. From Lemma 2-1 (ii) we have

$$|f(y)| \leq \frac{|f(x)| + |\beta|}{1 - |\overline{f(x)}||\beta|} .$$

Since $(A + t)/(1 + At)$ is increasing if $|A| \leq 1$, we get the result by replacing $|\beta|$ by $\sigma(x, y)$.

Lemma 2-5. If $f \in A$, $u = \mathrm{Re} f > 0$, then

(2-4) $$u(y) \geq \frac{1 - \sigma(x, y)}{1 + \sigma(x, y)}\, u(x) .$$

Proof. For real $t > 0$, let $g = e^{-tf}$, so that $|g| = e^{-tu} < 1$. From Lemma 2-4,

$$\begin{aligned} 1 - e^{-tu(y)} &= 1 - |g(y)| \\ &\geq 1 - \frac{\sigma(x, y) + |g(x)|}{1 + \sigma(x, y)\,|g(x)|} \\ &= \frac{(1 - \sigma(x, y))(1 - |g(x)|)}{(1 + \sigma(x, y)\,|g(x)|)} \\ &= \frac{1 - \sigma(x, y)}{1 + \sigma(x, y)e^{-tu(x)}} \cdot (1 - e^{-tu(x)}) \end{aligned}$$

Dividing both sides by t, and letting $t \downarrow 0$, we get

$$u(y) = -\frac{d}{dt} e^{-tu(y)}\Big|_{t=0} \geq \frac{1 - \sigma(x, y)}{1 + \sigma(x, y)}\, u(x)$$

The definition of $d(x, y)$ can be reformulated as follows:

$$\begin{aligned} d(x, y) &= \sup\{|\log u(x) - \log u(y)|:\ u \in B,\ u > 0\} \\ &= \inf\{\log a:\ a^{-1} \leq u(x)/u(y) \leq a,\ \text{all}\ u \in B,\ u > 0\} . \end{aligned}$$

provided we interpret the second line as $+\infty$ if no number a will work for all positive $u \in B$. The inequality (2-4) shows that $(1 + \sigma(x, y))/(1 - \sigma(x, y))$

is a number a which works for all $u > 0$, and hence that

(2-5) $$d(x, y) \leq \log \frac{1 + \sigma(x, y)}{1 - \sigma(x, y)} .$$

Theorem 2-6. $$d(x, y) = \log \frac{1 + \sigma(x, y)}{1 - \sigma(x, y)} .$$

Proof. If $d(x, y) = \infty$, then there is nothing more to prove in view of (2-5), so we assume that $d(x, y) < \infty$, and let $R(x, y) = \inf \{a: a^{-1} \leq u(x)/u(y) \leq a , \text{ all } u > 0\}$. We know that $R \leq (1 + \sigma)/(1 - \sigma)$, and we must prove equality. Let $f = u + i v \in A$, $\|f\| < 1$, $f(x) = 0$, and $f(y) = u(y) > 0$. Let $P = (1 + f)/(1 - f) \in A$. Then $\operatorname{Re} P = (1 - |f|^2)/|1 - f|^2 > 0$, $\operatorname{Re} P(x) = 1$, and so

$$\frac{\operatorname{Re} P(y)}{\operatorname{Re} P(x)} = \frac{1 + f(y)}{1 - f(y)} \leq R(x, y)$$

Since $(1 + t)/(1 - t)$ is increasing for $|t| < 1$, and $\sigma(x, y) = \sup |f(y)|$, $\|f\| < 1$, $f(x) = 0$, we have

$$\frac{1 + \sigma(x, y)}{1 - \sigma(x, y)} \leq R(x, y) .$$

Lemma 2-7. $\sigma(x, z) \leq \dfrac{\sigma(x, y) + \sigma(y, z)}{1 + \sigma(x, y)\sigma(y, z)} \leq \sigma(x, y) + \sigma(y, z)$, and consequently σ is a metric.

Proof. Let $f \in A$, $\|f\| < 1$, so that from 2-3 (with z for y) we have

$$|f(z)| \leq \frac{\sigma(x, z) + |f(x)|}{1 + \sigma(x, z)\,|f(x)|} .$$

If $f(y) = 0$, then $|f(x)| \leq \sigma(x, y)$, and since $(A + t)/(1 + At)$ is increasing if A $(= \sigma(x, z)) \leq 1$, we conclude that

$$|f(z)| \leq \frac{\sigma(x, z) + \sigma(x, y)}{1 + \sigma(x, z)\sigma(x, y)} .$$

Hence

$$\sigma(y, z) \leq \frac{\sigma(y, x) + \sigma(x, z)}{1 + \sigma(y, x)\sigma(x, z)} ,$$

which is the desired inequality with the letters permuted.

Theorem 2-8. $$\frac{4G(x, y)}{4 + G(x, y)^2} = \sigma(x, y)$$

Proof. From $0 \leq (2|\alpha| - |\alpha - \beta|)^2$ we conclude that for all α , β ,

$$4|\alpha||\alpha - \beta| \leq 4|\alpha|^2 + |\alpha - \beta|^2 .$$

We therefore have the following estimate if $|\alpha| \leq 1$

$$\begin{aligned} 4|1 - \bar{\alpha}\beta| &= 4|1 - |\alpha|^2 + \bar{\alpha}(\alpha - \beta)| \\ &\leq 4(1 - |\alpha|^2) + 4|\alpha||\alpha - \beta| \\ &\leq 4(1 - |\alpha|^2) + 4|\alpha|^2 + |\alpha - \beta|^2 \\ &= 4 + |\alpha - \beta|^2 \end{aligned}$$

From (2-2) we have for $\|f\| < 1$

$$4|f(x) - f(y)| \leq 4\sigma(x, y)|1 - \overline{f(x)}f(y)| .$$

and combining this with the inequality above gives

$$4|f(x) - f(y)| \leq \sigma(x, y)[4 + |f(x) - f(y)|^2] .$$

Since $4t/(4 + t^2)$ increases for $|t| \leq 2$, and $|f(x) - f(y)| \leq G(x, y)$ for $\|f\| < 1$, we have

$$\frac{4G(x, y)}{4 + G(x, y)^2} \leq \sigma(x, y)$$

To show the reverse inequality, for given $|\alpha| < 1$, let $\lambda = \alpha/(1 + \sqrt{1 - |\alpha|^2})$, so that $\lambda = r\alpha$ for some real number r . Then $|\lambda| < |\alpha| < 1$, and

$$\begin{aligned} |\lambda|\sqrt{1 - |\alpha|^2} &= |\alpha| - |\lambda| , \\ |\lambda|^2(1 - |\alpha|^2) &= |\alpha|^2 - 2|\alpha||\lambda| + |\lambda|^2 , \\ |\alpha|^2(1 + |\lambda|^2) &= 2|\alpha||\lambda| , \\ |\alpha| &= \frac{2|\lambda|}{1 + |\lambda|^2} . \end{aligned}$$

Since λ is a real multiple of α ,

(2-6) $$\alpha = \frac{2\lambda}{1 + |\lambda|^2} .$$

Computation shows that $$\frac{\lambda - \alpha}{1 - \bar{\lambda}\alpha} = -\lambda .$$

In other words, if $\phi(z) = (\lambda - z)/(1 - \bar{\lambda}z)$, then ϕ is the fractional linear transformation which sends α into $-\lambda$ and 0 into λ . This gives the maximum separation of α and 0 which can be induced by a fractional linear transformation.

Now let $f \in A$, $\|f\| < 1$, $f(x) = 0$, $f(y) = \alpha$, and let λ be as above. Let $g = \phi \circ f$; i.e., $g = (\lambda - f)/(1 - \bar{\lambda}f)$, so that $g \in A$, $\|g\| < 1$, $g(x) = \lambda$, and $g(y) = \phi(\alpha) = -\lambda$. Then, using (2-6)

$$\frac{4|g(x) - g(y)|}{4 + |g(x) - g(y)|^2} = \frac{2|\lambda|}{1 + |\lambda|^2} = |\alpha| .$$

Consequently, $|\alpha| \leq 4G(x, y)/(4 + G(x, y)^2)$, and hence $\sigma(x, y) \leq 4G(x, y)/(4 + G(x, y)^2)$.

<u>Theorem 2-9</u>. $d(x, y) = \log \frac{1 + \sigma(x, y)}{1 - \sigma(x, y)} = 2 \log \frac{2 + G(x, y)}{2 - G(x, y)}$.

<u>Proof</u>. The first inequality is Theorem 2-6. The identity

$$\frac{1 + \sigma(x, y)}{1 - \sigma(x, y)} = \left(\frac{2 + G(x, y)}{2 - G(x, y)}\right)^2 .$$

is immediate from Theorem 2-8.

<u>Notes</u>. Gleason was the first to discover that $G(x, y) < 2$ defines an equivalence relation [12]. Bishop used ideas similar to those involved in the definition of σ for his proof that $G(x, y) < 2$ defines an equivalence relation. The definition of d , and the fact that d and G give the same topology are shown in [4]. The identity (2-1) is due to König [14, 15].

3. GEOMETRIC DESCRIPTION OF PARTS IN T_B

We recall that if X is embedded (as evaluation functionals) in the dual space B' , with the w*-topology (w (B', B) topology), then the closed convex hull of X is the carrier space

$$T_B = \{F \in B': F(1) = \|F\| = 1\} .$$

Since T_B is a closed subset of the unit ball of B', T_B is a compact convex set. The function space B is isometric with the restrictions to T_B of all $w(B', B)$ continuous linear functionals on B'. We will therefore assume here that B is defined on all of T_B, and use x, y, z, etc. for points of T_B.

Definition 3-1. For any convex set K in a linear space, and points x, y of K, we say that the segment $[x, y]$ extends by $r > 0$ in K iff $x + r(x - y) \in K$ and $y + r(y - x) \in K$. We write $x \approx y$ iff $[x, y]$ extends in K by some positive r.

Lemma 3-2. If K is any convex set, and $[x, z]$ extends in K by a, $[z, y]$ extends in K by b, then $[x, y]$ extends in K by c, where $(1 + a^{-1})(1 + b^{-1}) = (1 + c^{-1})$. Consequently $\approx$ is an equivalence relation in K.

Proof. Let c be defined by the equation $(1 + a^{-1})(1 + b^{-1}) = (1 + c^{-1})$, so that

$$c = \frac{ab}{1 + a + b} .$$

If $\lambda = (1 + b)/(1 + a + b)$, $\mu = (1 + a)/(1 + a + b)$, then one can check that the following identities hold:

$$x + c(x - y) = \lambda[x + a(x - z)] + (1 - \lambda)[z + b(z - y)] ,$$
$$y + c(y - x) = \mu[y + b(y - z)] + (1 - \mu)[z + a(z - x)] .$$

The right sides above are convex combinations of points which are in K by hypothesis. Hence $[x, y]$ extends by the given value of c. The relation $\approx$ is clearly reflexive and symmetric, and is transitive by the argument above.

Definition 3-3. For x, y in a convex set K, and $x \approx y$,

$$D(x, y) = \inf \{\log (1 + \tfrac{1}{r}): [x, y] \text{ extends by } r \} .$$

Lemma 3-4. D is a metric on the equivalence classes of $\approx$ in any lineless convex set.

Proof. If $x \approx y$ and K does not contain a whole line, then $[x, y]$ extends by some $r > 0$ but not by arbitrarily large r if $x \neq y$. Therefore $D(x, y) > 0$ if $x \neq y$. If $[x, y]$ extends by a and $D(x, y) + \varepsilon > \log (1 + \frac{1}{a})$ and $[y, z]$ extends by b and $D(y, z) + \varepsilon > \log (1 + \frac{1}{b})$, then $[x, z]$ extends by c, and

$$\begin{aligned} D(x, z) &\leq \log (1 + \tfrac{1}{c}) \\ &= \log (1 + \tfrac{1}{a}) + \log (1 + \tfrac{1}{b}) \\ &< D(x, y) + D(y, z) + 2\varepsilon \end{aligned}$$

Consequently, D satisfies the triangle inequality, and since symmetry is obvious, D is a metric.

Theorem 3-5. If B is a function space on T_B, and $x, y \in T_B$, then $x \sim y$ iff $x \approx y$ and $d(x, y) = D(x, y)$.

Note. Both $\approx$ and D are purely geometric notions, independent of any class of functions. Hence, this theorem does give a purely geometric description of the Gleason parts of a function space. The Gleason parts of a function algebra A are the intersections of S_A with the parts of T_B, so this is also a sort of geometric description of the parts of S_A.

Proof. Assume first that $x, y \in T_B$ and that $x \not\approx y$; i.e., the segment $[x, y]$ does not extend, and to be definite we may assume that $[x, y]$ does not extend beyond x in T_B. Let z_n be the point of B' such that

(3-1) $$x = \frac{n - 1}{n} z_n + \frac{1}{n} y .$$

Then $z_n \notin T_B$, but $z_n(1) = x(1) = y(1)$. The point z_n can be separated from the compact convex set T_B by some $w(B', B)$-continuous linear functional on B'; i.e., by some $u_n \in B$. Therefore there is $u_n \in B$ such that $u_n(z_n) < 0$ and $u_n > 0$ on T_B. From (3-1) we conclude that $u_n(x) < \frac{1}{n} u_n(y)$ and hence $x \not\sim y$.

Now assume that $x \not\sim y$, and to be definite that $u(x)/u(y)$ is not bounded away from zero for u positive on T_B . Then there is no $z \in T_B$ such that $x \in (y, z)$, for if $x = \lambda y + (1 - \lambda)z$, $0 < \lambda < 1$, then

$$\frac{u(x)}{u(y)} = \frac{\lambda u(y) + (1 - \lambda)u(z)}{u(y)} \geq \lambda$$

for all positive u . Therefore if $x \not\sim y$, then $[x, y]$ does not extend in T_B

To see that $d = D$, note that $[x, y]$ extends by r means that $(1 + r)x - ry \in T_B$ and $(1 + r)y - rx \in T_B$. A point z is in T_B if and only if $u(z) \geq 0$ for all u positive on T_B by the Hahn-Banach Separation Theorem. Therefore $[x, y]$ extends by r , iff

$$(1 + r)u(x) - ru(y) > 0 \ ,$$

$$(1 + r)u(y) - ru(x) > 0 \ ,$$

for all $u \in B$ which are positive on T_B . Hence $[x, y]$ extends by r iff, for all $u > 0$,

$$(1 + \frac{1}{r})^{-1} < u(x)/u(y) < (1 + \frac{1}{r}) \ .$$

$$|\log u(x) - \log u(y)| < \log (1 + \frac{1}{r}) \ .$$

Since $d(x, y)$ is the supremum of the left side and D is the infinum of the right side, $D = d$.

From now on we supress D and use d for the part metric in a part of a function space, or for the part metric in any lineless convex set.

<u>Theorem 3-6</u>. If B is a function space on X , and Δ is a part, then d is complete on Δ .

<u>Note</u>. This implies that d is complete also on the part $\hat{\Delta}$ of T_B such that $\Delta = \hat{\Delta} \cap X$. Since a function $u \in B$ is positive on T_B if and only if u is positive on X , the part metric in Δ is the same as that in $\hat{\Delta}$. Consequently, any part Δ of X must be a d-closed subset of its parent part $\hat{\Delta}$ of T_B .

<u>Proof</u>. Let $\{x_n\}$ be a d-Cauchy sequence in Δ . Since X is compact, some

subsequence $\{x_{n_K}\}$ converges weakly:

$$u(x_{n_K}) \to u(x)$$

for all $u \in B$. Since

$$d(x, y) = \sup \{|\log u(x) - \log u(y)|: \ u > 0\} .$$

the sequences $\{\log u(x_n)\}$ are Cauchy uniformly for positive functions u. For fixed $u > 0$, $\log u(x_{n_K}) \to \log u(x)$ from the weak convergence. Since $\{\log u(x_n)\}$ is Cauchy uniformly in positive u, $\log u(x_n) \to \log u(x)$ uniformly for $u > 0$. Therefore $d(x_n, x) \to 0$.

We will give later a somewhat more general completeness theorem, for d defined geometrically on lineless convex sets, with no function space in view.

The next theorems are maximum theorems which are elementary consequences of the geometry of parts. The behavior described here is in striking contrast to Garnett's Theorem characterizing Gleason parts topologically. Garnett shows [11] that if K is any completely regular σ-compact topological space then there is a function algebra A such that S_A contains a part homeomorphic to K, and moreover so that every bounded continuous function on K is the restriction to K of some $f \in A$. Since there can obviously be local maxima in K, but no local maxima in S_A, the wild parts of S_A cannot be isolated in S_A.

<u>Theorem 3-7</u>. If B is a function space on T_B and for some $u \in B$, $u(x) = \max u(\Delta)$ where Δ is the part of T_B containing x, then u is constant on Δ.

<u>Proof</u>. If $y \sim x$ and $u(y) < u(x)$, then the segment $[y, x]$ extends beyond x to some $z \in \Delta$. Since each $u \in B$ is linear on T_B, $u(z) > u(x)$, which contradicts the assumption.

The theorem above is unnatural in that one does not deal with a function space on its carrier T_B, but on some very small subset $X = T_B$. The following corollary gives a version of the above which is easier to apply.

Corollary 3-8. If B is a function space on X, and $u(x) = \max u(X)$, then u is constant on the part of X containing x.

The corollary above implies the corresponding theorem for a function algebra A by considering $B = \mathrm{Re}\, A$. We state this result next and give a different proof involving σ.

Theorem 3-9. If A is a function algebra on X, and $f(x) = \|f\|$, then f is constant on the part containing x.

Proof. We may assume that $f(x) = \|f\| = 1$. If $f(y) \neq f(x)$, and $g = \frac{1}{2}(1 + f)$, then $|g(y)| < |g(x)| = \|g\| = 1$. Hence $g^n(y) \to 0$, $g^n(x) = 1$, and $\sigma(x, y) = 1$. That is, $f(y) \neq f(x) = \|f\|$ implies $y \not\sim x$.

Theorem 3-10. If B is a function space on T_B, and U is a neighborhood of z in T_B, and $u(z) = \max u(U)$, then u is constant on the part of T_B containing z.

Proof. Let $u(z) = \max u(U)$, where U is a basic $w(B', B)$ open neighborhood of z:

$$(4\text{-}2) \qquad U = \{x \in T_B\colon\ w_i(z) - \varepsilon < w_i(x) < w_i(z) + \varepsilon\ ,\ i = 1\ ,\ \ldots\ ,\ N\}\ .$$

Let $x \sim z$ and let Δ be the part of T_B containing x and z. The segment $[x, z]$ extends beyond z in T_B so there are points $y_1, y_2 \in T_B$ on the line of x and z, so that $z \in (y_1, y_2)$. If y_1 and y_2 are close enough to z, they will satisfy the N-inequalities defining U in (4-2); that is, $|w_i(z) - w_i(y_1)| < \varepsilon$ for $i = 1, \ldots, N$, and similarly for y_2. If $u(x) \neq u(z)$, then either $u(y_1)$ or $u(y_2)$ will be greater than $u(z)$, which is a contradiction.

We can restate the result above as follows: <u>a function $u \in B$ can have a strict local maximum in T_B only at a one-point part; i.e., at an extreme point.</u>

This result is not of any great interest unless one knows something of how the interesting space, X, fits in T_B.

Notes. The definition of parts and the part metric in any lineless convex is introduced in [9]. Completeness of d is shown in [3] in a more general setting than given here. (See Section 9).

4. REPRESENTING MEASURES AND PARTS IN T_C

If $C = C_R(\Gamma)$, then C' is the set of signed measures on Γ , and the carrier space T_C is the set of positive probability measures on Γ .

Let B be a function space on X , and Γ be the Silov boundary of B . Then B is isometric with $B|\Gamma$. The points of X (or T_B) are functionals in B' , and these functionals can be extended to functionals on $C = C_R(\Gamma)$, without increasing the norm. That is, for each $x \in T_B$ there is at least one $\mu \in T_C$ such that μ restricted to $B|\Gamma$ gives (evaluation at) x :

$$\int_\Gamma u d\mu = u(x) \tag{4-1}$$

for all $u \in B$.

Definition 4-1. Any measure $\mu \in T_C$ which satisfies (4-1) for all $u \in B$ is called a representing measure for x .

Lemma 4-2. For any points $x, y \in T_B$, $x \sim y$ iff there are representing measures μ_x, μ_y such that $\mu_x \sim \mu_y$ and $d(x, y) = d(\mu_x, \mu_y)$.

Note. We will use the same symbol d for the part metric of T_B and that of T_C .

Proof. If $\mu_x \sim \mu_y$, then there is a number a such that for every positive μ in $C_R(\Gamma)$;

$$\frac{1}{a} < \frac{\int u d\mu_x}{\int u d\mu_y} < a . \tag{4-2}$$

If u is in particular a function of B , then (4-2) implies that $a^{-1} < u(x)/u(y) < a$, so $x \sim y$. It is clear that $d(x, y) \leq d(\mu_x, \mu_y)$ for any two representing measures μ_x, μ_y for x and y .

Assume that $x \sim y$, and let $R = R(x, y)$ so that $d(x, y) = \log R$. Then $x_\circ = (1 + R)x - Ry \in T_B$ and $y_\circ = (1 + R)y + Rx \in T_B$ since T_B is compact.

Let $\mu_{x_\circ}, \mu_{y_\circ}$ be representing measures for $x_\circ, y_\circ$. Since

$$x = \frac{1+R}{1+2R} x_\circ + \frac{R}{1+2R} y_\circ$$

the measure

$$\mu_x = \frac{1+R}{1+2R} \mu_{x_\circ} + \frac{R}{1+2R} \mu_{y_\circ}$$

is a representing measure for x. Similarly,

$$\mu_y = \frac{R}{1+2R} \mu_{x_\circ} + \frac{1+R}{1+2R} \mu_{y_\circ}$$

is a representing measure for y. The segment $[\mu_x, \mu_y]$ therefore extends by R in T_C, to $\mu_{x_\circ}$ and $\mu_{y_\circ}$. Therefore $d(\mu_x, \mu_y) \leq \log R = d(x, y)$. Since the reverse inequality is automatic, we are done.

The above lemma shows that if $x \sim y$, then some part of T_C contains representing measures for both x and y. It is not generally true, as we show in the next section, that for any part Δ of T_B there is a part of T_C which contains representing measures for every point of Δ. We turn now to a description of the parts and part metric in the set T_C of probability measures on any compact space Γ.

If $\mu, \nu \in T_C$, and $\mu \sim \nu$, then for every positive continuous function $u \in C = C_R(\Gamma)$,

$$R(\mu, \nu)^{-1} < \frac{\int u d\mu}{\int u d\nu} < R(\mu, \nu) .$$

Letting u be an L_1-approximation the characterisitc function of a measurable set E, we see that μ and ν are mutually absolutely continuous, and that

$$R(\mu, \nu)^{-1} \leq \mu(E)/\nu(E) \leq R(\mu, \nu)$$

for all measurable sets E of positive measure. Therefore $\mu = g\nu$ where g is (essentially) bounded by $R(\mu, \nu)$ and bounded away from zero by $R(\mu, \nu)^{-1}$. Conversely, if $\mu = g\nu$ for a bounded measurable positive g which is bounded away from zero, it is clear that $\mu \sim \nu$. We state these observations in the following lemma.

Lemma 4-3. The part Π_μ of T_C containing μ consists of all probability measures of the form $g\mu$ for all bounded positive measurable g which are bounded away from zero.

By virtue of Lemma 4-3, the measures in a part Π_μ are in a one-to-one correspondence with a subset of $L_\infty(\mu)$.

Theorem 4-4. Convergence in the part metric d in Π_μ is equivalent to convergence of the corresponding Radon-Nikodym derivatives in the $L_\infty(\mu)$ norm metric.

Proof. For simplicity we replace the points $g\mu$ of Π_μ by the derivative g , and write $u(g)$ for $u(g\mu) = \int ug d\mu$ where $u \in C_R(\Gamma)$. We replace μ by the function 1. As we have seen $d(g_n, g) \to 0$ iff $u(g_n) \to u(g)$ uniformly for $u \in C_R(\Gamma)$, $u > 0$, $u(1) \leq 1$. (Here we have normalized the positive functions at the point $\mu = 1 \cdot \mu$.)

Suppose $g_n \geq g + \varepsilon$ on a set E_n with $\mu(E_n) > 0$. Let u_n be a continuous positive function which approximates $\chi_{E_n}/\mu(E_n)$ in $L_1(\mu)$, and which satisfies $\int u_n d = u_n(1) \leq 1$. Then

$$\begin{aligned} u_n(g_n) - u_n(g) &= \int u_n(g_n - g)d\mu \\ &\geq \varepsilon \qquad \text{(approximately)} \end{aligned}$$

Hence $u(g_n)$ does not converge to $u(g)$ uniformly for positive u with $u(1) \leq 1$ That is, $\|g_n - g\|_\infty \not\to 0$ implies $d(g_n, g) \not\to 0$, or d-convergence implies L_∞ convergence.

If $\|g_n - g\|_\infty \to 0$, then clearly

$$\begin{aligned} |u(g_n) - u(g)| &= |\int u(g_n - g)d\mu| \\ &\leq \|g_n - g\|_\infty \int u d\mu \\ &\leq \|g_n - g\|_\infty \end{aligned}$$

if $0 < u$ and $u(1) \leq 1$.

We have shown in Section 3-4 that d is complete in the parts of T_B (or T_C) . The L_∞ metric is not complete on a part Π_μ of T_C , since positive measurable functions can converge in L_∞ to a function g which is zero on a set of positive μ measure, and the measure $g\mu$ would not be in Π_μ .

Notes. Theorem 4-2 appears in [9]. The relation between the part metric in Π_μ and L_∞ is treated in [5], [3].

5. INNER PARTS AND THEIR ASSOCIATED NORMED SPACES

A convex body in a linear topological space has an innermost part; viz., its interior. Similarly, finite dimensional convex sets have an innermost part, which is their interior in the space they span. We can define the "inner part" of a convex set quite generally, independent of topology. We shall show in Section 6 that this idea is related to the possibility of choosing mutually absolutely continuous representing measures for the points in one part of a function space.

Definition 5-1. Let K be a convex set in a real linear space. Let K^i consist of all points $x \in K$ such that for every $y \in K$, the segment $[y, x]$ extends beyond x in K; i.e., $x + r(x - y) \in K$ for some positive r depending on y. If $K^i \neq 0$, K^i will be called the inner part of K.

Lemma 5-2. If $K^i \neq 0$, K^i is a part of K.

Proof. If $x \in K^i$ and $y \in K^i$, then $x \sim y$ by definition, so K^i is contained in a part. If $x \in K^i$ and $y \sim x$, then $[z, y]$ extends beyond y in K for every $z \in K$, as we showed in Section 3-1. (A two-dimensional picture makes this very clear.)

Example 5-3. There are convex sets with no inner part.

Proof. Let K be all probability measures on the unit interval. Suppose $\mu \in K^i$. For each $x \in [0, 1]$, let $\mathring{x}$ denote unit point mass at x. If there is an inner point μ, then $\mu + r_x(\mu - \mathring{x}) \in K$ for each $x \in [0, 1]$, and some $r_x > 0$. For this measure to be positive, $\mu\{x\}$ must be positive. This clearly cannot happen for uncountably many x, so $K^i = 0$.

Lemma 5-4. Let K be a convex set with non-empty inner part, and $0 \in K^i$. Then

$E = \cup\{tK: \; t > 0\}$ is the linear space generated by K, and K is radial at 0 in E.

Proof. Let $x \in tK$, and $y \in sK$, so that $\frac{1}{t}x$, $\frac{1}{s}y \in K$, for $t > 0$, $s > 0$. Then

$$\frac{t}{t+s} \cdot \frac{x}{t} + \frac{s}{t+s} \cdot \frac{y}{s} \in K ,$$

and $x + y \in (t + s)K$. If $x \in tK$, or $\frac{1}{t}x \in K$, then $[\frac{1}{t}x , 0]$ extends beyond 0 to $-\varepsilon x \in K$, since $0 \in K^i$. Hence $-x \in \frac{1}{\varepsilon}K$, or $-x \in E$ if $x \in E$. K is obviously radial at 0 in E.

For a convex set K which is radial at 0 in a linear space E, recall that the Minkowski functional for K is

$$p(x) = \inf \{r > 0: \; x \in rK\} ;$$

p is a subadditive and positive homogeneous on E. If

$$q(x) = \max[p(x) , p(-x)] ,$$

then q is the Minkowski functional for the symmetric convex $K \cap -K$, and q is a semi-norm. If K contains no line, q is a norm, which we will call the Minkowski norm for K.

The following can be thought of as an intrinsic characterization of those (lineless) convex sets which are convex bodies in <u>some</u> linear topological space. For lineless sets, it turns out that if K is a convex body in some l.t.s., then K is a convex body in a normed space.

Theorem 5-5. If K is a lineless convex, and $0 \in K^i$, and q is the Minkowski norm on $E = \cup\{tK: \; t > 0\}$, then K^i is the interior of K in E with the norm q.

Proof. For any convex K radial at zero, we have

$$\{x: \; p(x) < 1\} \subset K \subset \{x: \; p(x) < 1\} .$$

Since $[0, x]$ extends in K for all $x \in K^i$,

$$K^i = \{x: \ p(x) < 1\} .$$

We note that if p^i is the Minkowski functional for K^i, then $p^i = p$ on E. To see this first note that $p^i \geq p$ since $K^i \subset K$. However, if $x \in rC$, then $x \in sC^i$ for all $s > r$, so $p^i = p$. We will use p for the Minkowski functional of both K and K^i.

The following shows that p is q-continuous:

$$p(x) - p(x_\circ) \leq p(x - x_\circ) \leq q(x - x_\circ) .$$

Hence, $K^i = \{x: \ p(x) < 1\}$ is an open set in E, with norm q. Consequently, $K^i \subset K^\circ$, where K° is the interior of K. If $x \in K \sim K^i$, then $tx \notin K$ for all $t > 1$, and hence $x \notin K$. Therefore $K^i = K^\circ$.

<u>Theorem 5-6</u>. If K is a lineless open convex neighborhood of 0 in a linear topological space E, then K is one part, and the part metric d and the Minkowski norm q of E give the same topology in K.

<u>Proof</u>. Clearly K is one part, and as in the proof of Theorem 5-5, $K = \{x: \ p(x) < 1\}$. Let $x_\circ \in K$, so that $p(x_\circ) < 1$. If $q(x - x_\circ) < 1 - p(x_\circ)$, then

$$\begin{aligned} p(x) &= p(x - x_\circ + x_\circ) \\ &\leq p(x - x_\circ) + p(x_\circ) \\ &\leq q(x - x_\circ) + p(x_\circ) \\ &< 1 . \end{aligned}$$

In particular, $q(x - x_\circ) < 1 - p(x_\circ)$ implies $x \in K$. Since $q(x_\circ + r(x_\circ - x) - x_\circ) = rq(x_\circ - x)$, $x_\circ + r(x_\circ - x) \in K$ if $rq(x_\circ - x) < 1 - p(x_\circ)$ That is: $[x, x_\circ]$ extends beyond $x_\circ$ by r if

(5-1) $$r < (1 - p(x_\circ))/q(x_\circ - x) .$$

We also have $q(x + r(x - x_\circ) - x_\circ) = (1 + r)q(x - x_\circ)$, so $x + r(x - x_\circ) \in K$

if $(1 + r)q(x - x_0) < 1 - p(x_0)$. That is $[x_0, x]$ extends by r beyond x in K if

(5-2) $$r < \frac{1 - p(x_0) - q(x - x_0)}{q(x - x_0)} .$$

From (5-1) and (5-2) we conclude that $[x, x_0]$ extends by r in K for every r less than the right side of (5-2). As $q(x - x_0) \to 0$, $[x, x_0]$ extends by arbitrarily large r , or $d(x, x\ ) \to 0$.

To show that d-convergence implies q-convergence, assume that $x_0 \in K$, and that $[x, x_0]$ extends in K by (large) r , so that

$$p(x_0 + r(x_0 - x)) < 1 ,$$
$$p(x + r(x - x_0)) < 1 .$$

Notice that

(5-3) $$\begin{aligned} rp(x_0 - x) &= p(x_0 + r(x_0 - x) - x_0) \\ &\leq 1 + p(-x_o) . \end{aligned}$$

Similarly,

$$\begin{aligned} rp(x - x_0) &\leq 1 + p(-x) \\ &= 1 + p(x_0 - x - x_0) \\ &\leq 1 + p(x_0 - x) + p(-x_o) . \end{aligned}$$

Using (5-3), we get

(5-4) $$\begin{aligned} rp(x - x_0) &\leq 1 + \frac{1}{r}(1 + p(-x_0)) + p(-x_0) \\ &= (1 + \frac{1}{r})(1 + p(-x_o)) . \end{aligned}$$

From (5-3) and (5-4) we have that both $p(x_0 - x)$ and $p(x - x_0)$ are dominated by

$$\frac{1}{r}(1 + \frac{1}{r})(1 + p(-x_0)) .$$

Hence as $d(x, x_0) \to 0$ $(r \to \infty)$, $q(x - x_0) \to 0$.

Corollary 5-7. If P is a lineless part of any convex set K , and $0 \in P$, then the linear space generated by P is $E = \cup\{tP\colon\ t > 0\}$, and P is open in E

with the Minkowski norm q ; and d and q give the same topology in P .

Corollary 5-8. If K is a bounded open set in a normed linear space, then K has one part and d and the norm give the same topology in K .

Proof. We may assume without loss of generality that $0 \in K$, since translation changes neither the norm-distances nor d . The given norm and the Minkowski norm are equivalent. To see this note that 0 is in some open norm ball contained in $K \cap -K = \{x: q(x) < 1\}$. Since K is bounded, the q unit ball is also contained in some norm ball.

We will give in Section 9 some simple and natural sufficient conditions that the part metric be complete. The next theorem shows that if d is complete, the associated normed space is a Banach space.

Theorem 5-9. Let K be a lineless convex set, $0 \in K^i$, and $E = \cup\{tK: t > 0\}$ with the Minkowski norm q . If the part metric d of K^i is complete, then q is a complete norm.

Proof. Let $\{x_n\}$ be a q-Cauchy sequence. Since $|q(x_m) - q(x_n)| \leq q(x_m - x_n)$, $\{q(x_n)\}$ is convergent, and therefore a bounded sequence. Let $q(x_n) \leq B$. If $x_n' = x_n/2B$, then $p(x_n') \leq q(x_n') \leq M < 1$ and $\{x_n'\}$ converges if and only if $\{x_n\}$ does. Therefore without loss of generality we can assume that $p(x_n) \leq M < 1$, and in particular that $x_n \in K^i$ for all i .

In the proof of Theorem 5-7 (see (5-2)) we showed that $[x_n, x_m]$ extends in K by every r such

$$r < \frac{1 - p(x_n) - q(x_n - x_m)}{q(x_n - x_m)} .$$

Therefore $[x_n, x_m]$ extends by every r such that

$$r < \frac{1 - M - q(x_n - x_m)}{q(x_n - x_m)} .$$

For m , n large, the right side above is large, and hence $\{x_n\}$ is a d-Cauchy sequence. Let $d(x_n, x_0) \to 0$, where $x_0 \in K^i$. Since d and q give the same topology in K^i , $q(x_n, x_0) \to 0$, and q is complete.

The theorems above give a correspondence between inner parts of convex sets and the normed spaces generated by the convex sets. Since any part of any convex set is convex, and is an inner part in itself, we also have a correspondence between parts and the normed spaces generated by them (after translation to the origin). We now exploit this correspondence to show how the open mapping theorem for Banach spaces implies an open mapping theorem for an affine bijection of a one part convex set onto another.

Definition 5-10. Let K_1, K_2 be two convex sets, and f a mapping on K_1 to K_2. We call f affine iff f preserves convex combinations: $f(\lambda x + (1 - \lambda)y) = \lambda f(x) + (1 - \lambda)f(y)$ for all x, $y \in K_1$ and $\lambda \in [0, 1]$.

Lemma 5-11. If f is an affine map on K_1 to K_2, then $d_2(f(x), f(y)) \leq d_1(x, y)$, where d_i is the part metric of K_i.

Proof. Assume $[x, y]$ extends by r in K_1, so that $x_\circ = x + r(x - y) \in K_1$ and $y_\circ = y + r(y - x) \in K$. Then

$$x = \frac{1 + r}{1 + 2r} x_\circ + \frac{r}{1 + 2r} y_\circ ,$$

$$y = \frac{r}{1 + 2r} x_\circ + \frac{1 + r}{1 + 2r} y_\circ .$$

Consequently, $f(x)$, $f(y)$ are the same convex combinations of $f(x_\circ)$, $f(y_\circ) \in K_2$. Therefore $[f(x), f(y)]$ extends by r in K_2 to

$$f(x_\circ) = f(x) + r(f(x) - f(y)) ,$$

$$f(y_\circ) = f(y) + r(f(y) - f(x)) .$$

Corollary 5-12. If f is an affine map on K_1 to K_2, then f maps each part of K_1 into a part of K_2.

Theorem 5-13. Let K_1 be a one-part convex set and f an affine map on K_1 onto K_2 (so that K_2 is one part), and let d_1, d_2 be the corresponding part metrics. Then f is almost open if K_2 is d_2-complete, and f is open if K_1 and K_2 are both complete. (f is almost open if $\overline{f(U)}$ is a neighborhood of every point in $f(U)$).

Proof. Translation does not change the part metric, so we assume that $0 \in K_1$, $0 \in K_2$, and $f(0) = 0$. Let E_1, E_2 be the linear spaces generated by K_1, K_2, and let q_1, q_2 be their Minkowski norms. Then q_i is complete if d_i is.

The affine map f has a unique linear extension F mapping E_1 onto E_2 defined by $F(tx) = tf(x)$ for $x \in K_1$, $t > 0$. Since f is d_1, d_2 continuous on K_1, f is q_1, q_2 continuous on K_1, and therefore F is continuous. It follows that F is q_1, q_2 open by the standard Banach open mapping theorem. Since d_i open subsets of K_i are the same as q_i open sets $(i = 1, 2)$, it follows that f is d_1, d_2 open on K_1 onto K_2.

The following corollary is a consequence of the Bartle-Graves selection theorem for Banach spaces [1].

Corollary 5-14. Let K_1, K_2 be one-part convex sets with complete metrics d_1, d_2 and let f be an affine map on K_1 onto K_2. If g is a continuous map of a paracompact space T into K_2, then there is a continuous selection $\phi: T \to K_1$ such that $g = f \circ \phi$.

Proof. A map g into K_2 is d_2 continuous iff g is q_2 continuous. The Bartle-Graves theorem gives a selection function ϕ which is q_1 continuous and therefore also d_1 continuous.

Notes. The material of this section is developed in [3] and [2].

6. SELECTION OF MUTUALLY ABSOLUTELY CONTINUOUS REPRESENTING MEASURES

Let B be a function space on X, with Silov boundary Γ, and let $C = C_R(\Gamma)$. We consider the possibility of finding, for a given part Δ of X or T_B, a part Π_μ of T_C such that every point of Δ has a representing measure in Π_μ. That is, we want to find a measure μ so that each $x \in \Delta$ has a representing measure $g_x \mu$, where g_x is a bounded positive measurable function which is bounded away from zero. Sometimes such mutually absolutely continuous representing measures exist, and sometimes they do not. We begin with an example.

Let A be the cylinder algebra with centers identified (Section 1, Example 4). That is, X is $\overline{D} \times [0, 1]$, and A is the algebra of all continuous functions $f(z, t)$ on X such that $f(\cdot, t)$ is analytic on $|z| < 1$ and $f(0, t)$ is a constant. The set $\overline{D} \times [0, 1]$, with $\{0\} \times [0, 1]$ identified to a point, is one part Δ. The Silov boundary is $\{(z, t): |z| = 1\} = \Gamma$. We could alternatively consider the space $B = \text{Re } A$, or its closure, which is all functions $u(z, t)$ which are harmonic for $|z| < 1$, and have $u(0, t)$ constant.

Theorem 6-1. If $z_0 \neq 0$, and μ is a representing measure for (z_0, t_0), then the t_0-circle, $\Gamma_{t_0} = \{(z, t): t = t_0 ; |z| = 1\}$, has positive μ measure. Consequently, it is not possible to choose mutually absolutely continuous representing measures for points of the part Δ.

Proof. Suppose that Γ_{t_0} has zero μ-measure. Let S be a strip around Γ_{t_0} which has μ measure less than $\epsilon = |z_0|$; i.e. $S = \{(z, t): |z| = 1, t_0 - \delta < t < t_0 + \delta\}$ and $\mu(S) < |z_0|$. Let $f(z, t) = \phi(t) \cdot z$, where ϕ is a continuous real function, $0 \leq \phi(t) \leq 1$, $\phi(t_0) = 1$, and $\phi(t) \equiv 0$ if $t \notin (t_0 - \delta, t_0 + \delta)$. Clearly $f \in A$, and since μ represents (z_0, t_0), we have the contradiction

$$\begin{aligned} |f(z_0, t_0)| &= |z_0| \\ &= \left|\int_S \phi(t) z \, d\mu\right| \\ &\leq 1\mu(S) \\ &< |z_0| . \end{aligned}$$

Since any μ which represents (z_0, t_0) has positive mass on Γ_{t_0}, there must be many values of t such that $\mu(\Gamma_t) = 0$. It follows that there is no choice of mutually absolutely continuous measures.

Now let B be any function space, and Δ any non-trivial part. For $x \in \Delta$, we let P_x be the set of all representing measures (on Γ) for x. Each P_x is of course a w* compact convex subset of T_C, where $C = C_R(\Gamma)$. We show that if some P_x has an inner part, then all P_y do, for $y \sim x$, and all measures in these inner parts are in the same part of T_C.

Lemma 6-2. If $x \sim y$, and $R = R(x, y)$, then there are measures $\mu_x \in P_x$, $\mu_y \in P_y$ such that $P_x - \mu_x \subset R(P_y - \mu_y)$ and $P_y - \mu_y \subset R(P_x - \mu_x)$.

Proof. We pick μ_x, μ_y as in Lemma 4-2 so that

(6-1) $$R^{-1}\mu_x \leq \mu_y \leq R\mu_x .$$

If $\mu \in P_y$, then $\mu_x + R^{-1}(\mu - \mu_y) = (\mu_x - R^{-1}\mu_y) + R^{-1}\mu \geq 0$, and hence the measure is in P_x. That is, for all $\mu \in P_y$,

$$R^{-1}(\mu - \mu_y) + \mu_x \in P_x ,$$

or

$$R^{-1}(P_y - \mu_y) + \mu_x \subset P_x ,$$

or

$$P_y - \mu_y \subset R(P_x - \mu_x) .$$

The same argument shows that

$$P_x - \mu_x \subset R(P_y - \mu_y) .$$

Lemma 6-3. Let K_1 and K_2 be two convex sets, with $0 \in K_1 \cap K_2$, $K_1^i \neq 0$, and $cK_1 \subset K_2$, $cK_2 \subset K_1$ for some $c \leq 1$. Then $K_2^i \neq 0$ and $cK_1^i \subset K_2^i$, $cK_2^i \subset K_1^i$.

Proof. Let $x \in K_1^i$. To show $cx \in K_2^i$, we let $y \in K_2$, and show that $[y, cx]$ extends beyond cx in K_2. Since $cy \in K_1$, and $x \in K_1^i$, $[cy, x]$ extends beyond x in K_1 to some point $z \in K_1$:

$$x + r(x - cy) = z \in K_1 .$$

Since $cz \in K_2$, and $0 \in K_2$, $tz \in K_2$ if $0 \leq t \leq c$. We show that $[y, cx]$ extends past cx in K_2 to tz, where

$$t = \frac{c}{1 + r - rc^2} < c .$$

Specifically, the following identity can be verified by brute force:

$$cx + \left(\frac{rc^2}{1 + r - rc^2}\right)[cx - y] = tz$$
$$= \frac{c}{1 + r - rc^2}[x + r(x - cy)] \ .$$

A symmetric argument shows that $cK_2^i \subset K_1^i$.

We now have the following from Lemmas 6-2 and 6-3.

Lemma 6-4. If μ , ν are in the same part of P_x , then μ , ν are in the same part of T_C ; i.e., μ and ν are boundedly mutually absolutely continuous.

Proof. If $\mu \sim \nu$ in P_x , then there are μ_0 , $\nu_0 \in P_x$ such that $\mu = \lambda\mu_0 + (1 - \lambda)\nu_0$ and $\nu = \lambda\nu_0 + (1 - \lambda)\mu_0$. Then for any measurable set E

$$\frac{\mu(E)}{\nu(E)} = \frac{\lambda\mu_0(E) + (1 - \lambda)\nu_0(E)}{(1 - \lambda)\mu_0(E) + \lambda\nu_0(E)}$$
$$\leq \frac{\lambda}{1 - \lambda} + \frac{1 - \lambda}{\lambda} \ .$$

Theorem 6-5. If $P_x^i \neq 0$ for some x , then $P_y^i \neq 0$ for all $y \sim x$. All the inner parts P_x^i are subsets of some one part of T_C , so that if $\mu_x \in P_x^i$ for each x , the μ_x are all mutually absolutely continuous.

Proof. Assume $P_x^i \neq 0$ for some fixed x , and let y be any point equivalent to x . Pick $\mu_0 \in P_x$, $u_o \in P_y$ as in Lemma 6-2 so that $c(P_x - \mu_0) \subset P_y - \nu_0$ and $c(P_y - \nu_0) \subset P_x - \mu_0$, with $c^{-1} = R = R(x, y) > 1$. Clearly $(Py - \nu_0)^i = P_y^i - \nu_0$, so by Lemma 6-3, $c(P_y^i - \nu_0) \subset P_x^i - \mu_0$ and $c(P_x^i - \mu_0) \subset P_y^i - \nu_0$. Let $\mu_1 \in P_x^i$, and let $c(\mu_1 - \mu_0) = \nu_1 - \nu_0$ where $\nu_1 \in P_y^i$. Hence

$$\nu_1 = \nu_0 + c(\mu_1 - \mu_0) \in P_y^i \ .$$

Let $\gamma = \mu_1 - \mu_0$, so that $\gamma \in B^\perp$. Then

(6-2)
$$\mu_1 = \mu_0 + \gamma \in P_x^i \ ,$$
$$\nu_1 = \nu_0 + c\gamma \in P_y^i \ .$$

If

(6-3)
$$\mu_2 = \mu_0 + \tfrac{1}{2}c\gamma \ ,$$
$$\nu_2 = \nu_0 + \tfrac{1}{2}c\gamma \ ,$$

then $\mu_2 \in P_x$, $\nu_2 \in P_y$. Clearly $\mu_2 \sim \mu_1$ since $\mu_2 \in (\mu_0, \mu_1)$ and $\mu_1 \in P_x^i$, and similarly $\nu_2 \sim \nu_1$. Hence $\mu_2 \in P_x^i$ and $\nu_2 \in P_y^i$.

Now we show that $\mu_2 \sim \nu_2$. Suppose

$$\mu_2(E) = \mu_0(E) + \tfrac{1}{2}c\gamma(E) = 0 .$$

If $\gamma(E) < 0$, then $\mu_1(E) < 0$ which is impossible since $\mu_1 \in P_x$. Therefore, $\gamma(E) = \mu_0(E) = 0$. Since $\nu_0 \sim \mu_0$, $\nu_0(E) = 0$, and hence $\nu_2(E) = 0$. Therefore, μ_2 and ν_2 are mutually absolutely continuous. Now let E be a measurable set of positive μ_2 and ν_2 measure. From (6-2) we know that

$$(\mu_0 + c\gamma)(E) \geq 0 , \quad (\nu_0 + c\gamma)(E) \geq 0 ,$$

so that

$$c\gamma(E) \geq -\mu_0(E) ,$$
$$c\gamma(E) \geq -\nu_0(E) ,$$

and hence

$$\tfrac{1}{2}c\gamma(E) \geq -\tfrac{1}{2}\mu_0(E) ,$$
$$\tfrac{1}{2}c\gamma(E) \geq -\tfrac{1}{2}\nu_0(E) .$$

Therefore,

$$\frac{(\mu_0 + \frac{1}{2}c\gamma)(E)}{(\nu_0 + \frac{1}{2}c\gamma)(E)} = \frac{\mu_0(E)}{(\nu_0 + \frac{1}{2}c\gamma)(E)} + \frac{\frac{1}{2}c\gamma(E)}{(\nu_0 + \frac{1}{2}c\gamma)(E)}$$

$$\leq \frac{\mu_0(E)}{\frac{1}{2}\nu_0(E)} + 1$$

$$\leq \frac{2}{c} + 1$$

since $\mu_0 \leq c\nu_0$. This shows that $\mu_2 \leq (\frac{2}{c} + 1)\nu_2$, and the symmetric argument shows the other inequality. Hence $\mu_2 \sim \nu_2$.

What we have done is show that there are some points $\mu_2 \in P_x^i$, $\nu_2 \in P_y^i$ such that $\mu_2 \sim \nu_2$ (in T_c), and this is true for any x and y in the same part. Since all measures in P_x^i are equivalent, it follows that all measures in $\bigcup_x P_x^i$ are in one part of T_C .

Corollary 6-6. If $B^\perp$ = the real measures on Γ which are $\perp B$ is finite dimensional, then each P_x is finite dimensional, and hence has non-empty inner part. Therefore if $B^\perp$ is finite dimensional, equivalent representing measures for the points of any part can be found.

The last theorems give reasonable conditions under which there exist mutually absolutely continuous representing measures for all the points in some part. Next we use 4-4 and 5-14 to show that these representing measures, $\mu_x = g_x\mu$, can be chosen so that the Radon-Nikodym derivatives g_x vary continuously (in $L_\infty(\mu)$) with x . We will then show in the next section how the measures can be given by means of a kernel function, $g_x d\mu = Q(x, \theta)d\mu(\theta)$, in the separable case.

Theorem 6-7. Let B be a function space on T_B , and let Δ be a part of T_B . Suppose each x in Δ has a representing measure in the part Π_μ of T_C . Then one can choose representing measures $g_x\mu$ for $x \in \Delta$ so that $x \to g_x$ is a continuous map on Δ , d into $L_\infty(\mu)$ with the $L_\infty(\mu)$ norm.

Proof. Every measure $g\mu$ in Π_μ represents a point in Δ when restricted to B , since measures in the same part of T_C are clearly in the same part of T_B . The function f which restricts a measure in T_C to B is an affine continuous map on Π_μ , and f is onto Δ by hypothesis. It follows that f is open with respect to the respective part metrics of Π_μ and Δ . Therefore by 5-14 there is a continuous selection $g\colon \Delta \to \Pi_\mu$ so that $f \circ g$ is the identity map; i.e., $x \to g_x\mu$ is continuous on Δ into Π_μ , and $g_x\mu$ represents x . The fact that $g_x\mu$ is continuous in the part metric of Π_μ means that $x \to g_x$ is continuous in the $L_\infty(\mu)$ norm by 4-4.

Notes. The example of Theorem 6-1 and the fact that the sets P_x^i are all in one part Π_μ of T_C is contained in [13]. The continuous selection theorem (Theorem 6-7) is announced in [7].

7. INTEGRAL KERNELS

Let B be a function space on X . We have X divided into Gleason parts by the relation $x \sim y$ if $d(x, y) < \infty$, where d is the part metric defined by:

$$d(x, y) = \sup \{|\log u(x) - \log u(y)|: \ u \in B \ , \ u > 0\} .$$

As we saw in Section 3, $x \sim y$ iff there is a number a ($= (1 + \frac{1}{r})$, if $[x, y]$ extends by r in T_B) such that

$$a^{-1} < u(x)/u(y) < a \tag{7-1}$$

for all positive $u \in B$. We let $R(x, y)$ be the infinum of the numbers a which satisfy (7-1), so that $d(x, y) = \log R(x, y)$.

Let Δ be a fixed non-trivial part of X . There are three readily available topologies on Δ . We let $\mathcal{T}$ denote the topology that Δ has as a subset of X . Let $\mathcal{T}_d$ be the topology on Δ given by the part metric d . If we consider the points of X as evaluation functionals in the dual space B' , then X and hence Δ can be given the metric topology of the norm in B' . Identifying x with the evaluation-at-x functional, we have

$$||x - y|| = \sup \{|u(x) = u(y)|: \ u \in B \ , \ ||u|| \leq 1\} .$$

We let $\mathcal{T}'$ be this topology on Δ that Δ has as a subset of the dual space B' Otherwise described, the topology $\mathcal{T}_d$ is uniform convergence on the set of $\log u$, $u > 0$. The topology $\mathcal{T}'$ is uniform convergence on the unit ball of B , and the topology $\mathcal{T}$ is pointwise convergence on B ($x_\alpha \to x$ in $\mathcal{T}$ if and only if $u(x_\alpha) \to u(x)$ for all $u \in B$) .

Clearly $\mathcal{T}' \supset \mathcal{T}$, and we show next that $\mathcal{T}_d \supset \mathcal{T}'$.

<u>Lemma 7-1</u>. $\mathcal{T}_d \supset \mathcal{T}' \supset \mathcal{T}$.

<u>Proof</u>. Suppose that x_n , $x \in \Delta$, and $d(x_n, x) \to 0$. That is, $\log u(x_n) \to \log u(x)$ uniformly for $u \in B$, $u > 0$. If $v \in B$, $||v|| \leq 1$, then $1 \leq v + 2 \leq 3$. Hence, $|\log(v(x_n) + 2) - \log(v(x) + 2)| < \varepsilon$ if $n \geq N$,

where N depends on ε but not v. Therefore, taking exponentials, $|[v(x_n) + 2] - [v(x) + 2]| < \varepsilon \cdot e^3$ if $u \geq N$, and consequently $v(x_n) \to v(x)$ uniformly for $\|v\| \leq 1$. That is, $x_n \to x$ in $\mathcal{T}'$ if $d(x_n, x) \to 0$.

Definition 7-2. For a fixed point z in the part Δ, let

$$B^+(z) = \{u|\Delta: \ u \in B \ , \ u > 0 \ , \ u(z) = 1\} .$$

The set $B^+(z)$ is an abstract version of the set of positive harmonic functions on the unit disc, normalized to be one at a fixed point z.

Lemma 7-3. $\mathcal{T}_d$ is the weakest topology on Δ such that $B^+(z)$ is equicontinuous. Hence $B^+(z)$ is equicontinuous with respect to the given relativized topology $\mathcal{T}$ of X iff $\mathcal{T} = \mathcal{T}_d$.

Proof. Let $d(x_n, x) \to 0$, and let $u \in B^+(z)$. Then $R(x, z)^{-1} \leq u(x)/u(z) \leq R(x, z)$, or since $u(z) = 1$ for $u \in B^+(z)$, $R^{-1}(x, z) \leq u(x) \leq R(x, z)$ for $u \in B^+(z)$. Since $d(x_n, x) \to 0$, $R(x_n, x) \to 1$; that is, $u(x_n)/u(x) \to 1$ uniformly for $u > 0$. Since $u(x)$ is bounded for $u \in B^+(z)$, $u(x_n) \to u(x)$ uniformly for $u \in B^+(z)$. That is, $B^+(z)$ is equicontinuous (at each $x \in \Delta$) with respect to the topology $\mathcal{T}_d$.

Suppose conversely that $x_\alpha \to x$ in some topology in which $B^+(z)$ is equicontinuous, so that $u(x_\alpha) \to u(x)$ uniformly for $u \in B^+(z)$. Then $u(x_\alpha)/u(x) \to 1$ uniformly for $u \in B^+(z)$, since $R(x, z)^{-1} \leq u(x) \leq R(x, z)$ for $u \in B^+(z)$. Hence $R(x_\alpha, x) \to 1$, or $d(x_\alpha, x) \to 0$. That is, convergence in any topology in which $B^+(z)$ is equicontinuous implies convergence in $\mathcal{T}_d$.

Note that $\{u|\Delta: \ u \in B \ , \ u > 0 \ , \ u(z) \leq 1\}$ is equicontinuous if and only if $B^+(z)$ is, so this set can also be used to describe $\mathcal{T}_d$.

Next we prove an abstract version of Herglotz Theorem for positive harmonic functions.

Herglotz Theorem: For every positive harmonic function u on the open unit disc D there is a positive measure μ on the circle $\Gamma = \{z: \ |z| = 1\}$ such that

$$u(z) = \int_\Gamma P(z, \theta)d\mu(\theta)$$

where $P(z, \theta) = (1 - |z|^2)/|z - \theta|^2$ is the Poisson kernel. Conversely, every measure μ gives a harmonic function via the integral above.

If $u(0) = 1$, then the corresponding measure μ is clearly a probability measure $(\mu(\Gamma) = 1)$. The Poisson kernel functions $P(\cdot, \theta)$ are positive harmonic functions on D, and are the extreme points of the convex set of positive harmonic functions which are one at zero. Every positive harmonic function on D is the limit (uniform convergence on compact subsets of D) of functions in Re A where A is the disc algebra. These are the facts we generalize next to the setting of a general function space B, with a non-trivial part Δ replacing the disc D.

If Δ has the topology $\mathcal{T}_d$, so that $B^+(z)$ is equicontinuous, then $B^+(z)$ is pre-compact in the topology of uniform convergence on d-compact subsets of Δ (the u.c.c. topology on $B^+(z)$). The u.c.c. closure of $B^+(z)$ in $C_R(\Delta)$ is then a compact convex set of positive continuous functions which are 1 at z. If E is the closure of the set of extreme points of $\overline{B^+(z)}$, then by Choquet's representation theorem each $u \in \overline{B^+(z)}$ is the barycenter of a positive probability measure μ on E:

$$u = \int_E e\,d\mu(e) .$$

This means that for each $x \in \Delta$,

$$u(x) = \int_E e(x)\,d\mu(e) . \tag{7-1}$$

We write $\langle x, e\rangle = e(x)$, to emphasize that $e(x)$ is a function on $\Delta \times E$. If E has the u.c.c. topology, then $\langle x, e\rangle$ is jointly continuous on $\Delta \times E$, and (7-1) is an abstract Herglotz representation with a kernel $\langle x, e\rangle$. In the classical case the extreme functions e (the Poisson kernels $P(\cdot, \theta)$) have the additional nice property that they are identified with the topological boundary Γ of the disc D. We state the results proved above in the following theorem.

<u>Theorem 7-4</u>. If $\mathcal{T} = \mathcal{T}_d$ on a part Δ, or equivalently if $B^+(z)$ is equicontinuous with respect to $\mathcal{T}$, then for each $u \in \overline{B^+(z)}$ there is a probability

measure μ on the closure E of the extreme points of $\overline{B^+(z)}$ such that

$$u(x) = \int_E \langle x, e\rangle d\mu(e)$$

for all $x \in \Delta$. The kernel $\langle x, e\rangle$ is jointly continuous on $\Delta \times E$, if E has the u.c.c. topology.

The theorem above has at least two disadvantages as an integral representation theorem. The integral is over a very abstract "boundary", E, and the functions of B itself are not given in terms of their boundary values. The next theorem gives an integral representation which is somewhat superior, albeit at the expense of stronger hypotheses.

We note that the assumption of the next theorem that there is a continuous map $x \to g_x$ on a part Δ of X into a part Π_μ of T_C will be satisfied if B is finite dimensional, or a fortiori if $B|\Gamma$ is dense in $C_R(\Gamma)$. (See Theorem 6-7).

Theorem 7-5. Let X be a separable compact space, and B a function space on X with Silov boundary Γ and $\Delta = X \sim \Gamma$ one part. Suppose that there is a continuous map $x \to g_x$ on Δ, d to $L_\infty(\mu)$ so that $g_x\mu$ represents x. Suppose also that $B^+(z)$ is equicontinuous for some (all) $z \in \Delta$ (or equivalently that $\mathcal{T}_d = \mathcal{T}$). Then there is a jointly measurable function $Q(x, \theta)$ on $\Delta \times \Gamma$ such that $Q(\cdot, \theta)$ is continuous for each $\theta \in \Gamma$, and $Q(x, \cdot) = g_x$ a.e.μ, and hence

$$u(x) = \int_\Gamma u(\theta)Q(x, \theta)d\mu(\theta) \tag{7-2}$$

for all $x \in \Delta$, $u \in B$.

Proof. Let $\mu_0 = g_z\mu$, where z is the normalizing point of Δ. Then of course $\Pi_{\mu_0} = \Pi_\mu$, and $g_x\mu = \frac{g_x}{g_z}\mu_0$. The map $x \to g_x/g_z$ is again continuous. We will therefore assume without loss of generality that μ represents the normalizing point $z \in \Delta$.

Let Δ_0 be a countable dense subset of Δ, and let $z \in \Delta_0$. Let $Q(x, \theta) \equiv 1$ for $\theta \in \Gamma$. For each $x \in \Delta_0$, pick a measurable function (defined everywhere on Γ) $Q(x, \cdot)$ so that $Q(x, \cdot) = g_x$ a.e.μ .

We will use D for the part metric in T_C , and write $D(g, h)$ instead of $D(g\mu, h\mu)$ for simplicity. We also write $R(g, h)$ for $R(g\mu, h\mu)$; i.e., $D(g, h) = \log R(g, h)$.

For $g\mu$, $h\mu \in \Pi_\mu$, we have

$$R(g, h)^{-1} \leq g/h \leq R(g, h) \qquad \text{a.e.}\mu ,$$

and consequently

$$|g - h| \leq g(R(g, h) - 1) \qquad \text{a.e.}\mu . \tag{7-3}$$

For x , $y \in \Delta_0$,

$$R(g_x, g_y)^{-1} \leq \frac{Q(x, \theta)}{Q(y, \theta)} \leq R(g_x, g_y)$$

holds for all θ except in some set E_{xy} of measure zero. In particular, since $Q(z, \theta) \equiv 1$,

$$R(g_x, g_z)^{-1} \leq Q(x, \theta) \leq R(g_x, g_z)$$

holds except on E_{xz} , which has measure zero. Let E be the countable union of the sets E_{xy} for x , $y \in \Delta_0$. For $\theta \in E$, and $x \in \Delta_0$, redefine Q so that $Q(x, \theta) \equiv 1$. Then

$$R(g_x, g_y)^{-1} \leq \frac{Q(x, \theta)}{Q(y, \theta)} \leq R(g_x, g_y)$$

and

$$R(g_x, g_z)^{-1} \leq Q(x, \theta) \leq R(g_x, g_z)$$

hold for all $x \in \Delta_0$ and all $\theta \in \Gamma$. Of course $Q(x, \cdot)d\mu(\cdot)$ still represents x .

From (7-3) we now have the following inequality holding everywhere on Γ for x , $y \in \Delta_0$.

$$\begin{aligned} |Q(x, \theta) - Q(y, \theta)| &\leq Q(x, \theta)(R(g_x, g_y) - 1) \\ &\leq R(g_x, g_z)(R(g_x, g_y) - 1) \end{aligned} \tag{7-4}$$

Let $x \in \Delta$, $x_n \in \Delta_0$ and $d(x_n, x) \to 0$. Then $D(g_{x_n}, g_x) \to 0$, so $D(g_{x_n}, g_{x_m}) \to 0$ as $n, m \to \infty$, and $R(g_{x_n}, g_{x_m}) \to 1$. Clearly $R(g_{x_n}, g_z) \to R(g_x, g_z)$ as $n \to \infty$, so

(7-5) $$|Q(x_n, \theta) - Q(x_m, \theta)| \leq (R(g_x, g_z) + \varepsilon)(R(g_{x_n}, g_{x_m}) - 1)$$

for large n, m. Consequently $\{Q(x_n, \cdot)\}$ converges uniformly on Γ. The same kind of estimate as (7-5) shows the uniform limit of $Q(x_n, \cdot)$ does not depend on the sequence x_n in Δ_0 which converges to x. We therefore define $Q(x, \theta)$, for $x \in \Delta$, $\theta \in \Gamma$ to be the uniform limit of any sequence $Q(x_n, \cdot)$ with $x_n \in \Delta_0$ and $d(x_n, x) \to 0$. Clearly $Q(x,\cdot)$ is measurable on Γ. Since (7-4) now holds for all $x \in \Delta$, all $\theta \in \Gamma$, and $D(g_x, g_y)$ tends to zero as $d(x, y)$ does, Q is a continuous function of x for each fixed $\theta \in \Gamma$. We also have

$$\begin{aligned} u(x) &= \lim u(x_n) \\ &= \lim \int_\Gamma u(\theta)Q(x_n, \theta)d\mu(\theta) \\ &= \int_\Gamma u(\theta)Q(x, \theta)d\mu(\theta) \end{aligned}$$

for all $u \in B$, all $x \in \Delta$, so that $Q(x, \cdot)d\ (\cdot)$ does represent x.

Let $\hat{B}$ be the closure of $B|\Delta$ in the sense of uniform convergence on compact subsets of Δ. The space $\hat{B}$ is an abstract version of all "harmonic" functions on Δ. If $B|\Gamma$ has enough functions, then $Q(x, \theta)$ is "harmonic" in x for almost all θ, so Q can be redefined to "harmonic" in x for all θ without destroying its other kernel properties.

<u>Theorem 7-6</u>. If $B|\Gamma$ is dense in $L_1(\mu)$ in addition to the hypotheses of Theorem 7-5, then there is a kernel Q such that $Q(\cdot, \theta) \in B$ for all $\theta \in \Gamma$.

<u>Proof</u>. The proof is given in [8].

<u>Notes</u>. The topologies $\mathcal{T}$, $\mathcal{T}'$, $\mathcal{T}d$ were studied in [5], [8]. The abstract Herglotz theorem (Theorem 7-4) appears in [6].

8. GEOMETRIC PROPERTIES OF d AND A CONTINUOUS SELECTION THEOREM

We defined part and part metric in Section 3 for any lineless convex set K. Here we investigate the geometric properties of the part metric d, and give a continuous selection theorem.

In this section we assume that K is an arbitrary lineless convex. Recall that if $x \sim y$ (meaning $[x, y]$ extends in K by some $r > 0$), then

$$d(x, y) = \inf \{\log (1 + \tfrac{1}{r}): [x, y] \text{ extends by } r\} .$$

Lemma 8-1. Let $x, x', y, y' \in K$, and $0 \leq \lambda \leq 1$. If $[x, x']$ and $[y, y']$ extend by r, then $[\lambda x + (1 - \lambda)y, \lambda x' + (1 - \lambda)y']$ extends by r.

Proof. $(1 + r)[\lambda x + (1 - \lambda)y] - r[\lambda x' + (1 - \lambda)y']$

$$= \lambda[x + r(x - x')] + (1 - \lambda)[y + r(y - y')] .$$

Hence the extension beyond $\lambda x + (1 - \lambda)y$ is a convex combination of points which are in K by hypothesis. The extension on the other side is proved similarly.

Corollary 8-2.

(i) $d(\lambda x + (1 - \lambda)y, \lambda x' + (1 - \lambda)y') \leq \max [d(x, x'), d(y, y')]$

(ii) If S is convex, $S \subset K$, and $d(x, S) < \varepsilon$, $d(y, S) < \varepsilon$, then $d(\lambda x + (1 - \lambda)y, S) < \varepsilon$ for all $\lambda \in [0, 1]$.

(iii) If S is convex, $\{x: d(x, S) < \varepsilon\}$ is convex; d-balls are convex.

Note that in the corollary above S need not be in one part of K.

Lemma 8-3. If x, $y \in K$ and $\phi(\lambda) = \lambda x + (1 - \lambda)y$, then ϕ is d-continuous on $(0, 1)$. If $x \sim y$, then ϕ is d-continuous on $[0, 1]$.

Proof. We will prove the second assertion. The first then follows since the open interval (x, y) is contained in one part. Let $x \sim y$ and $z_\lambda = \lambda x + (1 - \lambda)y$. We show that $d(z, y) \to 0$ as $\lambda \to 0$. The following identities are easily checked:

$$y + r(y - x) = y + \frac{r}{\lambda}(y - z_\lambda) ,$$

$$x + r(x - y) = z_\lambda + \frac{1 - \lambda + r}{\lambda}(z_\lambda - y) .$$

Hence if $[x, y]$ extends by r, then $[y, z_\lambda]$ extends by the minimum of r/λ and $(1 - \lambda + r)/\lambda$. As $\lambda \to 0$, $[y, z_\lambda]$ extends by arbitrarily large values, and hence $d(y, z_\lambda) \to 0$.

Corollary 8-4. The parts of K are connected, and hence are components.

Theorem 8-5. If $\psi(\lambda, x, y) = \lambda x + (1 - \lambda)y$, then ψ is d-continuous on $[0, 1] \times P \times P$ into P, for every part P of K.

Proof. Let $x, x_0, y, y_0 \in K$, and $\lambda, \lambda_0 \in [0, 1]$. Then we have

$$\begin{aligned} d(\lambda x + (1 - \lambda)y, \lambda_0 x_0 + (1 - \lambda_0)y_0) &\le d(\lambda x + (1 - \lambda)y, \lambda x_0 + (1 - \lambda)y_0) \\ &\quad + d(\lambda x_0 + (1 - \lambda)y_0, \lambda_0 x_0 + (1 - \lambda_0)y_0) \\ &\le \max [d(x, x_0), d(y, y_0)] \\ &\quad + d(\lambda x_0 + (1 - \lambda)y_0, \lambda_0 x_0 + (1 - \lambda_0)y_o) . \end{aligned}$$

The second term approaches 0 as $\lambda \to \lambda_0$ by Lemma 8-3.

Next we have a version of Michael's continuous selection theorem for parts.

Theorem (E. A. Michael [16]). If T is a paracompact topological space, and ϕ is a lower semicontinuous map on T to the non-empty closed convex subsets of a Banach space Y, then there is a continuous selection function $f: T \to Y$ such that $f(t) \in \phi(t)$ for all $t \in T$.

We recall that ϕ is lower semi-continuous (l.s.c.) provided $\{x: \phi(x) \cap U \neq 0$ is open whenever U is open. If $\phi = F^{-1}$ for some onto function F, then ϕ is l.s.c. if and only if F is an open mapping. If g is a continuous onto map, and ϕ is l.s.c., then $\phi \circ g$ is l.s.c. Recall also that compact spaces and metric spaces are paracompact.

The theorem above remains true if we replace Y by a lineless convex set with complete part metric d.

Theorem 8-6. If T is a paracompact topological space, and ϕ is a lower semi-continuous map on T to the non-empty d-closed convex sets of a lineless convex Y with complete part metric d, then there is a continuous selection function $f\colon T \to Y$ such that $f(t) \in \phi(t)$ for all $t \in T$.

Proof. Michael's proof works verbatim, with the observation that convex combinations are continuous and the fact that the ε d-ball around a convex set is convex. The proof is written out in detail in [3].

The continuous selection Theorem (5-14) is also, of course, a corollary to this theorem.

Notes. The results of this section appear in [3].

9. COMPLETENESS OF THE PART METRIC

We have already seen that the part metric is complete on any part of a function space. In this section we give sufficient conditions for completeness of the part metric defined geometrically in any lineless convex set. Our setting will be a lineless convex set in a weak linear topological space:

Definition 9-1. A weak space is a Hausdorff linear topological space E with the $w(E, E')$ topology.

We observe that the convex sets in T_B we have considered so far have been subsets of the dual space B', with the w*-topology (i.e., the $w(B', B)$ topology). Since B is the dual of B' with the w*-topology, B' with the w*-topology is a weak space.

A closed convex set K in a weak space is an intersection of half-spaces:

$$K = \bigcap_{F \in K^+} \{x\colon \ F(x) \geq 0\} ,$$

where K^+ is the set of all continuous affine functions which are non-negative on K. That is, $x \in K$ if and only if $F(x) \geq 0$ for all $F \in K^+$.

Definition 9-2. A weak space E is complete iff there is $x \in E$ such that

$F(x_\alpha) \to F(x)$ for all $F \in E'$ whenever $\{F(x_\alpha)\}$ is a Cauchy net for all $F \in E'$.

Theorem 9-3. If K is a lineless closed convex set in a complete weak space, then

(9-1) $$d(x, y) = \sup \{|\log F(x) - \log F(y)| : F \in K^+ , \ F(x) > 0\} .$$

Proof. First notice that a function $F \in K^+$ is either identically zero on a part of K, or strictly positive on a part. If on the contrary $F(x) > 0$ and $F(y) = 0$, with $x \sim y$, then since the segment from x to y extends beyond y in K, F would have to be negative on K by linearity. Assume $x \sim y$ and let $K_\circ^+$ be those functions of K^+ which are strictly positive on the part containing x. The segment $[x, y]$ extends by r iff $(1 + r)x - ry \in K$ and $(1 + r)y - rx \in K$. That is, iff

$$(1 + r)F(x) \geq rF(y) ,$$

and

$$(1 + r)F(y) \geq rF(x)$$

for all $F \in K^+$, and therefore iff the strict inequalities hold for all $F \in K_\circ^+$. That is, $[x, y]$ extends by r iff

$$(1 + \tfrac{1}{r})^{-1} < F(x)/F(y) < (1 + \tfrac{1}{r})$$

for all $F \in K_\circ^+$. Taking logarithms shows that $[x, y]$ extends by r iff

$$|\log F(x) - \log F(y)| < \log (1 + \tfrac{1}{r})$$

for all $F \in K_\circ^+$. Therefore

$$\sup_{F \in K^+} |\log F(x) - \log F(y)| = \inf \log (1 + \tfrac{1}{r}) = d(x, y) .$$

Hence the formula is correct if $x \sim y$.

Now assume that $x \nsim y$ and that $y + \varepsilon(y - x) \notin K$ for every $\varepsilon > 0$. Then there is $F_\varepsilon \in K^+$ such that $F_\varepsilon(y + \varepsilon(y - x)) < 0$, or

$$(1 + \varepsilon)F_\varepsilon(y) < \varepsilon F_\varepsilon(x) ,$$

or

$$F_\varepsilon(x)/F_\varepsilon(y) < 1 + \frac{1}{\varepsilon} .$$

Therefore $\log F(x) - \log F(y)$ is not bounded as F runs over K^+, and the right side of (9-1) gives ∞ as it should if $x \not\sim y$ $(d(x, y) = \infty)$. If $F_\varepsilon(y) = 0$, we replace F_ε by $F_\varepsilon + \delta$ where δ is a small positive number, and make the same argument.

Lemma 9-4. If $\{x_\alpha\}$ is a net in one part of K, and $F(x_\alpha) \to F(x)$ for all $F \in K_o^+$, and $\{\log F(x_\alpha)\}$ is a Cauchy net uniformly for $F \in K_o^+$, then $d(x_\alpha, x) \to 0$.

Note: Here K_o^+ consists of the functions of K^+ which are strictly positive on the part containing the x_α.

Proof. The numbers $\log F(x_\alpha)$ converge to some number, depending on F, but converge uniformly in F. Since $F(x_\alpha) \to F(x)$ for each $F \in K_o^+$, $\log F(x_\alpha) \to \log F(x)$ uniformly in F. Therefore $d(x_\alpha, x) \to 0$ by Theorem 9-3.

We will use the following theorem of Choquet, which will be proved in the next section.

Theorem 9-5. If K is a complete lineless convex set in a weak space E, then each continuous affine function F on L can be written $F = F_1 - F_2$, where $F_i \in K^+$.

It is a consequence of this theorem that the weak topology in E is the weak topology induced by K^+, for any complete lineless convex K in E.

Theorem 9-6. If K is a complete lineless convex in a weak space E, then the part metric for K is complete.

Proof. Let $\{x_n\}$ be a d-Cauchy sequence in one part of K, and let K_o^+ be the affine functions positive on this part. We know that $\{\log F(x_n)\}$ is a Cauchy sequence uniformly for $F \in K_o^+$. Let $\phi(F)$ be the limit of $\log F(x_n)$. For fixed $F \in K_o^+$, $F(x_n)$ converges to something $(\exp \phi(F))$. Therefore, by Choquet's theorem, $F(x_n)$ converges for every continuous affine F. Since K is weakly complete by hypothesis, there is $x \in K$ such that $F(x_n) \to F(x)$ for all continuous affine F. Therefore $\log F(x_n) \to \log F(x)$ for $F \in K_o^+$, and uniformly in F since $\{\log F(x_n)\}$ is Cauchy uniformly in F. Therefore $d(x_n, x) \to 0$.

Theorem 9-7. A weakly compact lineless convex in a weak space is complete. A lineless cone with a compact base is complete.

Proof. If K is compact, and $\{x_n\}$ is a d-Cauchy sequence, then $x_{n_\alpha} \to x$ for some subnet $\{x_{n_\alpha}\}$. Therefore $F(x_{n_\alpha}) \to F(x)$ for all $F \in K_o^+$. Since $\{\log F(x_n)\}$ is convergent, it follows that $\log F(x_n) \to \log F(x)$ for each $F \in K_o^+$. Hence $F(x_n) \to F(x)$ for each $F \in K_o^+$, and therefore for each continuous functional. Therefore K is complete.

A convex set S is a base for a cone K provided that given $x \in K \sim \{0\}$, there is a unique $s \in S$ such that $x = ts$ for some positive t. Suppose S is weakly compact, and suppose that $\{F(x_\alpha)\}$ is a Cauchy net for each continuous functional F. Let $x_\alpha = t_\alpha s_\alpha$, where $s_\alpha \in S$. Let $s_{\alpha_\nu} \to s$ for some subnet $\{s_{\alpha_\nu}\}$ of $\{s_\alpha\}$. Then for all continuous functionals F,

$$F(x_{\alpha_\nu}) = t_{\alpha_\nu} F(s_{\alpha_\nu}),$$
$$t_{\alpha_\nu}^{-1} F(x_{\alpha_\nu}) = F(s_{\alpha_\nu}) \to F(s).$$

Pick F so that $F(s) > 0$; this is possible since the definition of base excluded the possibility that $0 \in S$. Since $F(x_{\alpha_\nu})$ converges, it follows that $t_{\alpha_\nu}^{-1} \to t^{-1}$. Therefore $F(x_{\alpha_\nu}) \to F(ts)$ for all F, and hence $F(x_\alpha) \to F(ts)$ for all F.

Notes. The results of this section appear in [3].

10. LINEAR FUNCTIONALS AS DIFFERENCES OF POSITIVE FUNCTIONALS

In this section we give a proof of the theorem of Choquet [10].

Lemma 10-1. If E is a complete weak space, then E is isomorphic to the algebraic dual, E'^*, of E', with the $w(E'^*, E')$ topology.

<u>Proof</u>. For $s \in D$, let $\hat{x}$ be the element of E'^* defined by $\hat{x}(F) = F(x)$, $F \in E'$. The mapping $x \to \hat{x}$ is an isomorphism since E is a weak space, and E' separates E . Convergence in E is by definition pointwise convergence on E' , and this is also what convergence in $w(E'^*, E')$ means:

$$x_n \to x \quad \text{iff} \quad F(x_n) \to F(x) \quad \text{all} \quad F \in E' ,$$
$$\hat{x}_n \to \hat{x} \quad \text{iff} \quad \hat{x}_n(F) \to \hat{x}(F) \quad \text{all} \quad F \in E' .$$

Therefore E is isomorphic to a subspace $\hat{E} = \{\hat{x}: \ x \in E\}$ of the algebraic dual of E' .

We show that E is dense in E'^* , and then conclude that $E = E'^*$ since E (and hence $\hat{E}$) is complete.

If $\phi \in E'^*$, then a basic neighborhood U of ϕ is determined by $F_1, \ldots, F_n \in E'$ and $\varepsilon > 0$: $U = \{\psi \in E'^*: \ |\psi(F_i) - \phi(F_i)| < \varepsilon, \ i = 1, \ldots, n\}$ We show that there is $x \in E$ so that $\phi(F_i) = F_i(x) = \hat{x}(F_i)$. We may assume without loss of generality that $F_1, \ldots, F_n$ are independent. Pick $x_1 \in E$ such that $F_1(x_1) \neq 0$, and then a number c_1 so that $F_1(c_1x_1) = \phi(F_1)$. The null space of F_1 is not contained in the null space of F_2 , since F_1 and F_2 are independent. Hence there is $x_2 \in E$ such that $F_1(x_2) = 0$ and $F_2(x_2) \neq 0$. Pick c_2 so that

$$F_2(c_1x_1 + c_2x_2) = \phi(F_2)$$
$$F_1(c_1x_1 + c_2x_2) = F_1(c_1x_1) = \phi(F_1) .$$

Similarly, the null space of F_3 does not contain the intersection of the null spaces of F_1 and F_2 . So we choose x_3 so that $F_3(x_3) \neq 0$, $F_1(x_3) = F_2(x_3) = 0$, and then c_3 so that

$$F_3(c_1x_1 + c_2x_2 + c_3x_3) = \phi(F_3)$$
$$F_2(c_1x_1 + c_2x_2 + c_3x_3) = \phi(F_2)$$
$$F_1(c_1x_1 + c_2x_2 + c_3x_3) = \phi(F_1) .$$

Etc.

Since there is x so that $x(F_i) = \phi(F_i)$, $i = 1$, ... , n , E is dense. Since E, and hence $\hat{E}$, is complete, $\hat{E} = E'^*$.

Lemma 10-2. If E is a complete weak space, and B is an algebraic basis for E', then E is isomorphic to R^B, with the product topology (R is the real line).

Proof. We will show that $E'^* = R^B$, where the equality means isomorphism and homeomorphism of E'^* with the $w(E'^*, E')$ topology and R^B with the product topology.

The linear functionals ϕ on E' are in a one-to-one correspondence with all the real functions on the basis B. This correspondence is also clearly an isomorphism of E'^* and R^B with pointwise addition. Pointwise convergence of functionals on B is equivalent to pointwise convergence on E'^*, which is just finite linear combinations of elements of B. The product topology in R^B is also just pointwise convergence on B.

Lemma 10-3. If E is a complete weak space, then every closed subspace M of E has a topological supplement N; i.e., $E = M \oplus N$, and $m_\alpha + n_\alpha \to m + n$ if and only if $m_\alpha \to m$ and $n_\alpha \to n$.

Proof. Let M be closed in E, so M is a complete weak space. Let A be an algebraic basis for M'. Extend, by the Hahn-Banach theorem, each $F \in A$ to a functional $F_\circ \in E'$. If $A_\circ$ be the set of extensions, then $A_\circ$ is also an independent subset of E'^*. Extend $A_\circ$ to a set $A_\circ \cup B$ which forms a basis for E'^*. Then $M = R^A = R^{A_\circ}$ and $E = R^{A_\circ \cup B} = R^{A_\circ} \times R^B$. Therefore $N = R^B$ is a topological supplement of $M = R^{A_\circ}$.

Theorem 10-4. Let $K_\circ$ be a complete convex set in a weak l.t.s. E. Then there is a topological isomorphism $\phi\colon E \to R^I \times R^J$ such that $\phi(E)$ is dense and a translation of $\phi(K_\circ)$ is $R^I \times A$, where A is a closed convex set in R^J_+; i.e., for some $g \in R^I \times R^J$, $\phi(K_\circ) + g = R^I \times A$ and $A \subset R^J_+$. If $K_\circ$ is lineless, then E is dense in R^J and a translate of $K_\circ$ is isomorphic to $A \subset R^J_+$.

Proof. Let $\tilde{E}$ be the completion of E. Let $a \in K_\circ$, and $K = K_\circ - a$, so that K is a closed convex set containing the origin. Let $M = \{x \in K\colon Rx \subset K\}$, so that M consists of all lines through 0 which are in K. Intuitively, M is the largest subspace of $\bar{E}$ which is contained in K, so that K can contain no whole line in dimensions complementary to M. Note that $x \in M$ if and only if $rx \in K$ for all $r \in \mathbb{R}$.

We show first that M is a closed subspace of $\bar{E}$, and that $K + M = K$.

It is easy to check that M is a subspace of $\bar{E}$, and M is closed since K is. Clearly $K \subset K + M$, since $0 \in M$. Let $x = y + m$, with $y \in K$, $m \in M$,

$$\begin{aligned} x &= y + m \\ &= \lim_{n\to\infty} [\tfrac{n-1}{n}y + \tfrac{1}{n}(nm)] . \end{aligned}$$

Since $y \in K$, and $nm \in K$, and K is closed, $x \in K$.

Since K is a closed convex set, K can be written as an intersection of half-spaces:

$$K = \bigcap_{F \in S} \{x \in K:\ F(x) \leq \alpha_F\} ,$$

where S is some subset of E'. Since $0 \in K$, $\alpha_F \geq 0$ for all $F \in S$. If $F(x) = 0$ for all $f \in S$, then clearly $x \in K \subset M$. If $x \in M$, then $tx \in K$ for all $t \in \mathbb{R}$, and hence $F(tx) \leq \alpha_F$ for all t. Consequently, we have shown that

(10-1) $$M = \{x:\ F(x) = 0 \ \text{ all } \ F \in S\} .$$

Now let N be a topological supplement of M in E: $E = M \oplus N$. Let $K_N = K \cap N$. Then we assert that $K = M + K_N$.

Clearly $M + K_N \subset M + K = K$. Let $y \in K$, and $y = m + n$ for $m \in M$, $n \in N$. Then for $F \in S$, $F(y) = F(m) + F(n) = F(n)$. Therefore, $F(n) \leq \alpha_F$ whenever $m + n \in K$, so $n \in K$, and $K = M + K_N$.

Next we show that K_N is closed, convex, and lineless. Since both K and N are closed and convex, $K_N = K \cap N$ is obviously also closed and convex. If there is some line in K_N, then for some y, z, $y + tz \in K_N$ for all $t \in \mathbb{R}$. Then

$$\frac{1}{n}(y + ntz) + \frac{n-1}{n} \cdot 0 \in K_n$$

for all n . Since K_N is closed, $tz \in K_N \subset K$ for all t , and consequently $z \in M$. But we also know that $1 \cdot z \in K_N \subset N$. Since $M \cap N = \{0\}$, $z = 0$, and there is no line in K_N .

The set K_N can be written as an intersection of half-spaces in N :

$$(10\text{-}2) \qquad K_N = \bigcap_{F \in T} \left\{x \in N: \ F(x) \geq \alpha_F \right\}.$$

Since $0 \in K_N$, $\alpha_F \leq 0$ for each $F \in T$. We show that for any such subset T of N' , T spans N' . Let

$$H = \bigcap_{F \in T} \{x \in N: \ F(x) = 0\} ,$$

so that $H \subset K_N$. If T does not span N' , then there is a proper subset $T_\circ \subset T$ such that $T_\circ \cup I$ is a base for N' for some non-empty set $I \subset N'$. By Lemma 10-2 $N = R^{T_\circ \cup I} = R^{T_\circ} \times R^I$ (We will forego mention of the isomorphism ϕ , and identify our spaces with their images in the canonical product of lines). We know that $x \in H$ if and only if $F(x) = 0$ for all $F \in T$. This is equivalent to $F(x) = 0$ for all $F \in T_\circ$, since $T_\circ$ spans T . Since H , considered as embedded in $R^{T_\circ} \times R^I$, consists of all functions on $T_\circ \cup I$ which are zero on $T_\circ$, there are lines in H if $I \neq 0$. But $H \subset K_N$, which is lineless. Therefore $I = 0$, and T must span N' .

Now let $T_\circ \subset T$, where T satisfies (10-2),and $T_\circ$ is a basis for N' . Then

$$K_N \subset \bigcap_{F \in T_\circ} \{x \in N: \ F(x) \geq \alpha_F\} .$$

Now we have

$$E = M \oplus N = R^I \oplus R^{T_\circ} = R^I \times R^{T_\circ} ,$$
$$K = M + K_N = R^I \times K_N ,$$

and $K_N \subset R_F^{T_\circ}$, where $R_F = [\alpha_F, \infty]$. If we translate K by the function which is zero on the index set I , and equal to $-\alpha_F$ on $T_\circ$, then the translate is a set of form $R^I \times A$, where A is a closed convex set in $R_+^{T_\circ}$. Of course K is already a translate of our original set $K_\circ$. Note also that $K_\circ$ remains unchanged in the completion of E, since K_0 is complete.

<u>Corollary 10-5</u>. If K is a complete lineless convex in a weak space E, then every continuous linear functional on E is the difference of continuous affine functions which are positive on K.

<u>Proof</u>. Let the completion $\bar{E}$ be R^J, and let $K + g \subset R_+^J$. A linear functional Φ on R^J is continuous if and only if $\Phi(h) = c_1 h(F_1) + \ldots + c_n h(F_n)$ for $F_1, \ldots, F_n \in J$. Let $\Phi_1(h) = \Sigma c_i h(F_i)$ for those i with $c_i \geq 0$, and $\Phi_2(h) = \Sigma {-c_i} h(F_i)$ for those i with $c_i < 0$. Then $\Phi = \Phi_1 - \Phi_2$ and $\Phi_i \geq 0$ on $R_+^J \supset K + g$. Let $\Phi_1' = \Phi_1 + c$, $\Phi_2' = \Phi_2 + c$ where $c \geq |\Phi_i(g)|$. Then $\Phi = \Phi_1' - \Phi_2'$, and $\Phi_i' \geq 0$ on K.

BIBLIOGRAPHY

1. R. G. Bartle and L. M. Graves, Mappings between function spaces, Trans. Am. Math. Soc., Vol. 72 (1952), pp 400-413.

2. Heinz Bauer, An open mapping theorem for convex sets with only one part, to appear.

3. Heinz Bauer and H. S. Bear, The part metric in convex sets, Pacific J. Math., Vol. 30 (1969), pp 15-34.

4. H. S. Bear, A geometric characterization of Gleason parts, Proc. Am. Math. Soc., Vol. 16 (1965), pp 407-412.

5. H. S. Bear, Continuous subparts for function spaces, Function Algebras, Scott, Foresman and Co., 1966.

6. H. S. Bear, The integral representation of functions on parts, Illinois J. Math., Vol. 10 (1966), pp 49-55.

7. H. S. Bear, Continuous selection of representing measures, Bull. Am. Math. Soc., to appear.

8. H. S. Bear and Bertram Walsh, Integral kernel for one-part function spaces, Pacific J. Math., Vol. 23 (1967), pp 209-215.

9. H. S. Bear and M. L. Weiss, An instrinsic metric for parts, Proc. Am. Math. Soc., Vol. 18 (1967), pp 812-817.

10. G. Choquet, Ensembles et cônes convexes faiblement complets, C. R. Academie Sci., Vol. 254 (1962), pp 1908-1910; 2123-2125.

11. John Garnett, A topological characterization of Gleason parts, Pacific J. Math., Vol. 20 (1967), pp 59-63.

12. A. M. Gleason, Function algebras, Seminar on analytic functions, Vol. 2 Institute for Advanced Study, Princeton, 1957.

13. N. V. Harkova, Generalized Poisson formula (Russian), Vestnik Moskov. Univ., Ser. I Mat. Meh., Vol. 22 (1967), pp 25-30.

14. Heinz König, Zur abstrakten Theorie der analytischen Funktionen. II, Math. Annalen, Vol. 163 (1966), pp 9-17.

15. Heinz König, On the Gleason and Harnack metrics for uniform algebras, Proc. Am. Math. Soc., Vol. 22 (1969), pp 100-101.

16. E. Michael, Selected selection theorems, Am. Math. Monthly, Vol. 43 (1956), pp 233-238.

Offsetdruck: Julius Beltz, Weinheim/Bergstr

MIX
Papier aus verantwortungsvollen Quellen
Paper from responsible sources
FSC® C105338

If you have any concerns about our products,
you can contact us on
ProductSafety@springernature.com

In case Publisher is established outside the EU,
the EU authorized representative is:
Springer Nature Customer Service Center GmbH
Europaplatz 3, 69115 Heidelberg, Germany

Printed by Libri Plureos GmbH
in Hamburg, Germany